how to
know the

aquatic
insects

The **Pictured Key Nature Series** has been published since 1944 by the Wm. C. Brown Company. The series was initiated in 1937 by the late Dr. H. E. Jaques, Professor Emeritus of Biology at Iowa Wesleyan University. Dr. Jaques' dedication to the interest of nature lovers in every walk of life has resulted in the prominent place this series fills for all who wonder **"How to Know."**

John F. Bamrick and Edward T. Cawley
Consulting Editors

The Pictured Key Nature Series

How to Know the
AQUATIC INSECTS, Lehmkuhl
AQUATIC PLANTS, Prescott
BEETLES, Arnett-Downie-Jaques, Second Edition
BUTTERFLIES, Ehrlich
ECONOMIC PLANTS, Jaques, Second Edition
FALL FLOWERS, Cuthbert
FERNS AND FERN ALLIES, Mickel
FRESHWATER ALGAE, Prescott, Third Edition
FRESHWATER FISHES, Eddy-Underhill, Third Edition
GILLED MUSHROOMS, Smith-Smith-Weber
GRASSES, Pohl, Third Edition
IMMATURE INSECTS, Chu
INSECTS, Bland-Jaques, Third Edition
LAND BIRDS, Jaques
LICHENS, Hale, Second Edition
LIVING THINGS, Jaques, Second Edition
MAMMALS, Booth, Third Edition
MARINE ISOPOD CRUSTACEANS, Schultz
MITES AND TICKS, McDaniel
MOSSES AND LIVERWORTS, Conard-Redfearn, Third Edition
NON-GILLED FLESHY FUNGI, Smith-Smith
PLANT FAMILIES, Jaques
POLLEN AND SPORES, Kapp
PROTOZOA, Jahn, Bovee, Jahn, Third Edition
SEAWEEDS, Abbott-Dawson, Second Edition
SEED PLANTS, Cronquist

SPIDERS, Kaston, Third Edition
SPRING FLOWERS, Cuthbert, Second Edition
TREMATODES, Schell
TREES, Miller-Jaques, Third Edition
TRUE BUGS, Slater-Baranowski
WATER BIRDS, Jaques-Ollivier
WEEDS, Wilkinson-Jaques, Third Edition
WESTERN TREES, Baerg, Second Edition

how to
know the
aquatic
insects

Dennis M. Lehmkuhl
University of Saskatchewan

The Pictured Key Nature Series
Wm. C. Brown Company Publishers
Dubuque, Iowa

Copyright © 1979 by Wm. C. Brown Company Publishers

Library of Congress Catalog Card Number: 78–55761

ISBN 0–697–04767–9 (Paper)
ISBN 0–697–04766–0 (Cloth)

Printed in the United States of America
10 9 8 7 6 5 4 3

This book is dedicated to my past teachers.

Contents

Preface

Although insects are basically land animals in their structure and evolution, many thousands of species in about a hundred families of a dozen orders are found in the streams, rivers, lakes, bogs, springs, and on the seashores of North America. This invasion of water has given rise to a spectacular array of behavioral and morphological adaptations which have fascinated biologists for centuries. In recent years there has been renewed and intensified interest in aquatic insects because they have been recognized as essential members of ecological communities and the natural environment. Throughout North America, hundreds of surveys are taking place investigating the inhabitants of aquatic ecosystems and the possible effects new developments will have on them.

Despite this attention, there are innumerable cases where even the most basic facts of taxonomy and biology remain unknown, thus leaving the possibility of new discoveries for anyone who turns their attention to aquatic insects.

Freshwater investigations have been hampered by the problems of identifying the great variety of organisms encountered. Both naturalists and professional biologists often find that workable and up-to-date keys and guidebooks are not available. At the time of this writing,* the most recent comprehensive keys for North America are Pennak (1953), Usinger (1956), and Edmundson (1959). Not only has much changed since these books were written, but no single recent volume covers both adults and immature stages of all the aquatic insect families of North America. Several books are necessary to cover the field. This book attempts to fill the gap. It grew out of a key which I devised for use in an ecologically oriented field course. The book was prepared in particular for ecologists, students, naturalists, and those specialists who want general information on groups other than their own. It was written in the spirit of Lindroth (1969, P. XVII), that the objective is to communicate, and that taxonomy should not be a secret code for a handful of conspirators. Only time will tell if this attempt has been successful.

Many keys use "natural" characters in the couplets, thus imparting phylogenetic information. This key is decidedly artificial, and any character which "works" has been used. I have referred extensively to every published work available to me, but the keys presented, on the whole, differ considerably from any other source.

The over 600 drawings arranged in nearly 300 figures should aid in using the keys. Most drawings are diagrammatic; others are

*Merritt and Cummins (1978) has since been published (see General References).

recognition or habitus drawings. They were prepared either to supplement the text and convey a clear meaning, or to present a general impression of the appearance of the group. Therefore, species names are not given with the drawings.

This book is regarded as a practical tool and is meant to be a working companion in the field and in the laboratory. It has two aims. One is to provide a quick and accurate means of identifying families and distinctive genera. The family is the standard level for many teaching and survey purposes. It is hoped that from this point of view the book is a self-contained and independent unit. Second, it can be used as an introduction to the specialized literature. General information can be found in the present book; monographs or papers dealing with species level taxonomy or biology can be located by using the section on references.

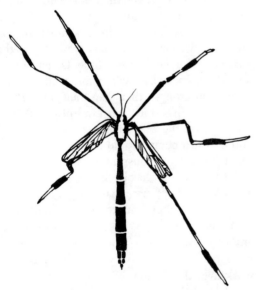

Any book of this type could not exist without the cumulative published results of all those who have worked in the field in the past. These persons number in the hundreds, and those whose work has been especially consulted are included in the section on references. It was necessary to omit many.

Collections on which this book is based began (unwittingly) about 15 years ago, particularly in the vicinities of the University of Montana, Missoula, Oregon State University, Corvallis, and the University of Saskatchewan, Saskatoon. Many additional trips covered at least in a cursory way most regions of the continent. For loan of additional specimens, I thank Dr. N. H. Anderson of Oregon State University, and Dr. W. P. McCafferty of Purdue University.

Most of the diagrams and figures were prepared by me from specimens in my collection. Numerous other drawings are unpublished originals by the late A. R. Brooks (Fig. 18B, C; Fig. 19C-H; Fig. 21A-D; Fig. 30A; Fig. 31A-C, F-H; Fig. 37B; Fig. 42; Fig. 53; Fig. 74A; Fig. 115A, B, E; Fig. 119B; Fig. 155A, B; Fig. 182A-K; Fig. 183A-C; Fig. 184; Fig. 185; Fig. 186; Fig. 187A, B; Fig. 188; Fig. 191A, B; Fig. 195; Fig. 197; Fig. 199B-N). While these are often styled after drawings published elsewhere, examination shows that they differ substantially from other sources and they are credited to Brooks. Several figures are from Brooks and Kelton (1967, Hemiptera references) and are used with permission (Figs. 17D, F-I; Fig. 167; Fig. 171B; Fig. 173). Fig. 148A is by Lloyd Dosdall, and Fig. 151 is redrawn from an original by Dosdall. Several unnumbered figures in the preface are by my wife, Bundy Suzanne.

Parts of the text and keys have been read or tested by the following individuals at the University of Saskatchewan, and their comments were very much appreciated: Lloyd Dosdall, Collembola, Plecoptera, Hemiptera; John Lawrence, Key to Orders and Ephemeroptera; Peter Mason, Diptera and Odonata; Douglas H. Smith, Trichoptera and Coleoptera. Dr. J. D. Shorthouse of Laurentian University, Sudbury, Ontario, read the introductory and appendix sections. Dr. N. H. Anderson of Oregon State University read the entire manuscript and made innumerable valuable comments and criticisms. His prompt attention to a flood of manuscript made it possible to meet a tight schedule and deadline. Errors and imperfections which remain are of course my responsibility.

As always, I thank my wife, Bundy Suzanne, for understanding the foibles of a biologist, not only during the preparation of this manuscript, but over the years.

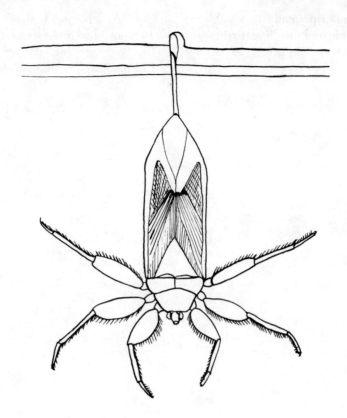

Introduction to
Aquatic Insects

THE DIVERSITY AND NUMBERS OF AQUATIC INSECTS

Aquatic insects are a varied group with one thing in common—a dependence on a water habitat at some stage in the life cycle. There is scarcely a natural aquatic habitat that some species does not occupy. They are found in salt lakes and thermal springs, in water-filled footprints and large lakes, in mountain torrents and stagnant swamps. It is probable that in any given area the size of a state or province, a minimum of 1500 species can be found. Estimates place the number of aquatic species at about 3% of the total insects, or about 25,000-30,000 world-wide (Cheng, 1976). The figures in Table 1, which include my rough percentage estimates in some cases, place the number of species for North America north of Mexico at about 6,500. (Most of the numbers are from Borror, DeLong, and Triplehorn, 1976.) This estimate indicates that about 7% of North American insects are aquatic or semiaquatic (based on a total of about 91,000 species for North America). Elsewhere, I have seen unpublished estimates that there are 6,000 species of aquatic insects in Canada alone, and that about 50% (rather than 10%) of all Diptera larvae are aquatic. This would give a much higher total for North America. Not only is information for an accurate count unavailable, but the definition of an aquatic insect allows wide interpretation. Many so-called aquatic species are obligatory shore dwellers but never are actually submerged in water. Including or excluding semiaquatic forms will alter the estimate considerably. This book follows a very broad definition of an aquatic insect, and semiaquatic species are included.

HABITATS OF AQUATIC INSECTS

There are few ecological niches in the freshwater environment that are not occupied by aquatic insects. In lakes and ponds the open water is inhabited by water boatmen, backswimmers, beetle larvae, and phantom midges (Fig. 1A). The surface film is occupied by whirlygig beetles, water striders, and Collembola, while in tangles of submerged vegetation there is usually an abundance of predatory damselflies and dragonflies as well as herbivorous mayflies. Bottom mud and gravel substrates may be occupied by burrowing mayflies and midge larvae. Sprawling on or crawling over the surface of the substrate are different species of mayflies, dragonflies, and beetles. Various adults occur on shore mud and in vegetation.

Table 1 Orders Containing Aquatic Insects

Order	Aquatic Stage (excluding egg)	North America North of Mexico	
		Approximate Number of Aquatic Species	Non-aquatic Species *(exceptions occur)
Collembola (Springtails)	All	25	300 total
Ephemeroptera (Mayflies)	Nymphs	625	None
Odonata (Damselflies & Dragonflies)	Nymphs	425	None
Orthoptera (Grasshoppers & Crickets)	All	5	1000
Plecoptera (Stoneflies)	Nymphs	425	None
Hemiptera (True Bugs)	All	400	4,600
Neuroptera (Lacewings)	Larvae	5	300
Megaloptera (Dobsonflies & Alderflies)	Larvae	40	None
Coleoptera (Beetles)	All	1000	30,000 total
Trichoptera (Caddisflies)	Larvae	1000	*None
Lepidoptera (Moths)	Larvae	50	10,000 total
Diptera (True Flies)	Larvae	10% of total? (see text)	17,000 total
Hymenoptera (Wasps)	Larvae, parasites	0.5% of total?	17,000 total
		approx. 6500 total	

Lakes and ponds are far behind in comparison to the insect fauna of streams and rivers (Fig. 1B). Highly modified nymphs and larvae of half a dozen families filter food particles from the passing current. Flattened nymphs live under rocks in torrents, while streamlined forms are at home exposed to strong currents. Tangled roots, calm backwaters, and the depths of the substrate each have characteristic insect inhabitants.

There are many less obvious habitats for aquatic insects. Some species are restricted to water-filled tree holes or the water in pitcher plants. Some are found only in the mud and under moss in the vicinity of springs and seep areas. Others occupy temporary or seasonal habitats which are dry a significant portion of the year. Salt lakes, thermal springs, brackish water, and even the substrate many feet below the surface water of streams has a characteristic fauna. Truly marine forms are discussed in a later section.

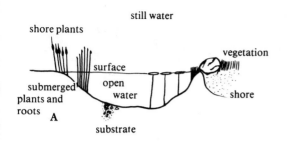

Figure 1   Habitats in still and flowing water.

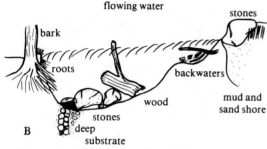

2E), and Sialidae (Megaloptera) deposit neat rows of eggs on leaves overhanging water (Fig. 2D). This is only a sample of egg laying behaviour.

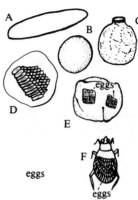

Figure 2   Insect eggs. A. Odonata; B. Ephemeroptera; C. Plecoptera; D. *Sialis* (Megaloptera); E. *Baetis* (Ephemeroptera); F. Belostomatidae (Hemiptera).

THE LIFE CYCLE OF INSECTS

Insect life cycles follow three basic patterns (Table 2) and any or all stages of the cycle may be aquatic. All insects hatch from eggs (Fig. 2). Eggs of aquatic insects and their mode of deposition are extremely varied. Some adults simply drop loose masses of individual eggs into the water and they are dispersed passively (many stoneflies, some mayflies and caddisflies). Many females have a sharp or elongate ovipositor at the end of the abdomen and eggs are laid individually in punctures in plant leaves or stems (Odonata, some Coleoptera and others). A few insects carry masses of eggs on their own bodies until they hatch (Hemiptera, Fig. 2F). Chironomidae (Diptera) and many caddisflies lay eggs in protective gelatin-like masses either in or out of the water. *Baetis* (Ephemeroptera) females crawl under the water in rapid currents to lay eggs in quarter-inch square patches on rocks (Fig.

Eggs may hatch immediately or they may require a temperature sequence to trigger them to develop. In harsh conditions (e.g. temporary ponds or the north), over half of the year, including the unfavorable period, may be passed as an inactive or diapausing egg. (See references, ecology section.)

Primitive insects such as Collembola are referred to as Ametabolous, or have no metamorphosis, since newly hatched individuals resemble mature forms and there is no definite externally visible transition from immature to adult. Collembola is the only example of this type among aquatic insects (Table 2).

Hemimetabolous or Paurometabolous insects (incomplete metamorphosis) have an immature stage, usually called a nymph or naiad (Fig. 3) which hatches from the egg and transforms to the adult stage in a single undelayed moult. Plecoptera, Ephemeroptera, Hemiptera, Odonata, and Orthoptera undergo this type. Nymphs assume a number of forms and occur in a variety of habitats, but they usually have wing buds, long antennae, and segmented tarsi. Nymphs hatch from the egg and

Table 2 Types of Metamorphosis in Insects

A. egg → miniature adult → growth and maturity→adult (Ametabolous; Collembola)
B. egg → nymph → adult
   (Hemimetabolous or with incomplete metamorphosis; Odonata, Plecoptera, Ephemeroptera, Orthoptera, Hemiptera)
C. egg → larva → pupa → adult
   (Holometabolous or with complete metamorphosis; Trichoptera, Lepidoptera, Neuroptera, Megaloptera, Hymenoptera, Diptera, Coleoptera).

grow through a number of moults without greatly changing in appearance. At the final moult a winged and usually sexually mature adult emerges. Nymphs may require from a few months to several years to complete their growth.

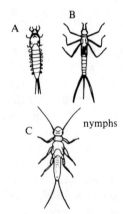

Figure 3 A. Ephemeroptera; B. Odonata; C. Plecoptera.

Complete metamorphosis (holometabolous) includes a larva (Fig. 4) plus the usually non-active pupa (Figs. 5, 22). The pupa bridges the gap between the active larval and adult stages (Fig. 6). This type of metamorphosis is found in Trichoptera, Lepidoptera, Coleoptera, Neuroptera, Megaloptera, Hymenoptera, and Diptera.

The term larva is here restricted to the growing and active stages of holometabolous insects (some authors do not use the term nymph, referring to all immatures as larvae).

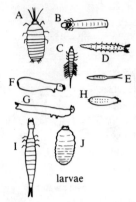

Figure 4 A. Neuroptera; B. Trichoptera; C. Megaloptera; D-H. Diptera; I, J. Coleoptera.

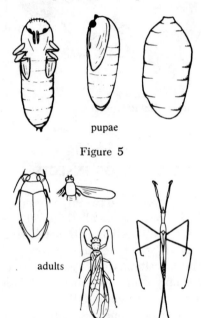

Figure 5

Figure 6

Larvae lack wing buds, usually have short antennae, have unsegmented tarsi, and are found in an extreme variety of shapes and sizes (Fig. 4). Larvae, after hatching from the egg, usually do not change in appearance other than for size, but when drastic body changes do occur, it is referred to as hypermetamorphosis (e.g. Hydroptilidae, Trichoptera). The last larval instar prepares a cell, digs a burrow, or spins a cocoon. Internal and external changes take place during the pupal stage, and after a time, the adult emerges.

Almost all insects reproduce by sexual reproduction, with mating and egg production taking place in the adult stage. A number of cases of parthenogenesis (hatching of eggs without being fertilized by the male) are known, as well as some cases of ovoviviparity (eggs develop in the body of the female, a type of live birth). Gynandromorph specimens (which are abnormalities of two sexes in one individual) are seen occasionally, but these individuals are not viable.

Two further points should be mentioned in this section. Modern insect physiologists are revising the classic concepts of life cycle classifications as given in Table 2. However, since most biological and taxonomic literature of the past uses the old concepts and names, they will be used in this book. The second point is that almost every aquatic biologist has an opinion about the correct use of the terms nymph, naiad, and larva. Some insist on using only *larva,* and essentially banishing the other words from the language. Others will argue for the use of all three. I use the word *nymph* for immature stages of those groups which do not have a pupal stage, and *larva* for the active immature stages of the groups which have a pupa (as in Table 2).

THE AQUATIC STAGES OF INSECTS

Some entire Orders are strictly aquatic in the immature stage (with the inevitable odd exception) while the adult stages in these Orders are typical terrestrial insects. These include the hemimetabolous Plecoptera, Ephemeroptera, and Odonata and the holometabolous Trichoptera and Megaloptera.

Other Orders are primarily terrestrial and only a small portion of the species are aquatic. The aquatic representatives may be an entire family [e.g. Mosquitoes (Culicidae), Predacious Water Beetles (Dytiscidae), and Water Boatman (Corixidae)] or the representatives may be odd species with strictly terrestrial relatives. Parasite wasps, members of many Diptera families (*Atherix,* Rhagionidae), and the aquatic Lepidoptera are examples. There are less common situations where the adult stage is aquatic and the larva is terrestrial (Hydraenidae, Coleoptera). Many beetles and true bugs come closest to being absolutely aquatic with all stages living in the water and seldom entering the air. This is possible because of highly adapted and efficient respiratory systems and the fact that the wings are modified into hard or leathery structures that are not a hinderance under water.

Many Coleoptera, Hemiptera, Orthoptera, and adult Diptera are restricted to muddy shores and sand beaches but only in this way are they aquatic. Collembola, some adult Diptera, many true bugs, and some beetles skate or hop over the surface of water and, like the shore inhabitants, are never truly submerged. Many beetles and true bugs enter or leave the water at will in the adult stage, usually feeding in the water and leaving for breeding and dispersal.

ECOLOGY OF AQUATIC INSECTS

While this book takes the approach that the insects themselves are the objects of interest, there are many situations where aquatic insects are studied as part of a larger system. Insects may serve as model organisms which illustrate a basic biological principle. Examples of such areas are the study of predator-prey relationships, trophic (food) relationships, modes of adaptation to extreme habitats such as temporary pools, downstream drift of organisms in streams, or diapause and its ecological ramifications. Detailed discussion is beyond the scope of this book, but in such studies the identification and general biology of the insect involved is an integral part of the overall project. A number of papers dealing with such studies are listed in the references, Ecology section.

THE ORIGIN OF AQUATIC INSECTS

While a considerable amount of information from fossils has accumulated on the origin of insects as a group, the complete story of the origin of aquatic insects is not known. It is generally agreed that insects are land animals. The annelid-like ancestor of insects was terrestrial, probably living in soil and leaf litter.

None of the primitive wingless insects which exist today is truly aquatic (Collembola are surface-film dwellers). The most ancient of the still existing winged orders, Ephemeroptera, is strictly aquatic in the nymphal stage. The other orders which are strictly aquatic in the immature stage, Odonata, Plecoptera, Megaloptera (Neuroptera), and Trichoptera, also appeared very early in the fossil record (Table 3). In a word, it seems that very soon after the origin of insects from a terrestrial annelid-like ancestor, and after the evolution of wings, many whole orders became aquatic in the immature stages. In later periods, there were numerous smaller invasions of fresh water, giving rise to single species or families of insects which are aquatic.

Table 3   Sequence of appearance of aquatic insect Orders in the fossil record
Years ago (millions)

Years ago (millions)	Era/Period	
0	Cenozoic Era (age of mammals)	no new aquatic insect Orders
70	Cretaceous (age of reptiles)	Lepidoptera
135	Jurassic	no new aquatic insect Orders
180	Triassic (Dinosaurs)	Diptera, Hymenoptera, Trichoptera
225	Permian	Odonata, Plecoptera, Hemiptera, Neuroptera, Coleoptera (Megaloptera)
270	Upper Carboniferous (Coal forests)	Ephemeroptera
	Lower Carboniferous	details unknown concerning insects
350 400	Devonian (age of fishes)	First wingless insects

Marine Insects

In order to round out the coverage of insects that live in water, it seems only mildly inappropriate to include at this point a brief discussion of marine insects. The sea, after all, is a "water" environment. Of the million or more species of insects in the world, only a few hundred are truly sea animals. Only about half a dozen species have conquered the open ocean. The sea appears to be the single major habitat where insects have not been successful. No one has proposed a completely satisfactory explanation of this failure, although one can fall back on the opinion of a world renowned entomologist whose answer is simply that insects and their remote ancestors, as land animals, never had anything to do with the sea, so why should they go back to it. This seems to be a clever remark but not an explanation. The same could be said for mammals, but they are successful in the sea, so why not insects? Other opinions are that some undetermined weak point in the life cycle or physiology prevents insects from coping with salt water. Another is that the multitude of invertebrates already in the sea provides too much competition for the newcomers from the land.

It is interesting to compare the physiological environment of freshwater and the sea. The problems faced by organisms living in freshwater and in salt water are in many ways opposite. Water, either salty or fresh, is of course taken in with food. Also, there is the problem of respiratory surfaces. These surfaces must be permeable so that the insect can take in oxygen and release carbon dioxide. This permeability means that water also passes through, thus upsetting the internal balance of the insect. In freshwater, the tendency is for the insect to gain water and dilute the body fluids. Freshwater insects must have efficient systems of eliminating water but retaining the correct balance of necessary salts and ions. It is the opposite in saltwater, where the salt concentration in some cases exceeds that of the body fluids of the insect. The direction of movement is to loose water from the body but to gain salt in the body, a potentially fatal situation. Marine insects, if they are actually in contact with salt water and not surface or shore dwellers, must have an efficient system of retaining water and excreting salt. In this respect living in salt water is similar to living on dry land. It is not peculiar, therefore, that relatively few freshwater insects have become truly marine, and that many truly marine insects (especially Coleoptera) are specialized forms of terrestrial families.

The overview of marine insects which follows is based mainly on Cheng (1976). Most marine families can be keyed in this book.

Many groups, such as Ephemeroptera, Plecoptera, Odonata, Neuroptera, and Megaloptera apparently do not have marine representatives.

Hemiptera are among the most successful marine insects, especially as shore or surface dwellers. About 5 species of *Halobates* in the family Gerridae have truly conquered the open ocean and can be found hundreds of miles at sea where they live on the surface. An additional hundred or so species of a dozen genera are associated with the surface of salt water along beaches and shores. Many are questionably marine in the physiological sense. These families include Hermatobatidae, Mesoveliidae, Veliidae, Gerridae, Hydrometridae, and Macroveliidae. The last 5 are covered in keys in this book. Hermatobatidae consists of about 10 species of exclusively marine insects. They are widespread near tropical islands in the Indian and Pacific Oceans. The tarsal claws are terminal on the middle and hind legs but inserted before the tips of the tarsi on the front legs, and the head is 3 times as wide as long from dorsal view. Among shore bugs, about 60 species in 20 genera of the families Gelastocoridae, Ochteridae, Omaniidae, and Saldidae (the latter with over 3/4 of the genera and species) are found in salt marshes, on reefs and on marine shores. Omaniidae are found in the region from east Africa to India and Australia. The other families are covered in the keys. In addition, about a dozen genera and over 50 species (worldwide) of Corixidae are well known to inhabit either inland saline water or the coastal marine environment. Many of these are basically freshwater forms which have a wide range of tolerance and are able to occupy a wide range of habitats.

Trichoptera also have truly marine representatives. There are no marine species known from North America but a species of *Limnephilus* is found in saltmarsh pools in Britain. In New Zealand and south Australia, several marine species are known which belong to the family Philanisidae = Chathamiidae in the genera *Philanisus* and *Chathamia*. Larvae crawl over corals and feed on algae. Cases are usually made of fragments of calcareous algae.

Like Hemiptera, a number of Diptera are associated with salt water. Many mosquito species breed in brackish water and saltmarshes. Nine genera are commonly found in tidal areas. Seven of these genera also have freshwater species. A similar situation exists for Ceratopogonidae. Three genera and many species of these biting midges breed in saline environments. The non-biting midges, Chironomidae, are also well represented in brackish and marine waters, with about a dozen species in half a dozen genera being truly marine in North America. Several species of Tabanidae, Tipulidae, and Dolichopodidae breed in coastal salt marsh areas. Ephydridae are often called brine flies because they are found in inland saline lakes as well as marine habitats. All the above Diptera are included in keys in this book. Not included are the seaweed flies, Coelopidae, which breed in piles of kelp and seaweed and are characterized by small to medium size, a bristly or hairy and rather flat body, and short antennae and stout legs.

While most freshwater aquatic beetle families have representatives which occasionally enter marine water, the majority of truly marine beetles are specialized members of terrestrial families. Several marine families are discussed in the keys. Families known to have marine representatives, sometimes only opportunistically and sometimes only as beach inhabitants, include Carabidae, Hydrophilidae, Hydraenidae, Limnichidae, Melyridae, Salpingidae, Tenebrionidae, Rhizophagidae, Chrysomelidae, Curculionidae, Haliplidae, Dytiscidae, Anthicidae, Heteroceridae and Staphylinidae.

Non-Insects
Frequently Encountered

Since the aquatic entomologist will encounter a number of non-insect arthropods (joint-legged animals) which could cause confusion, some of the most common types are shown in Fig. 7. Most are crustaceans, the group which includes crabs and lobsters. Exceptions are Fig. 7H, the water mites, which are relatives of spiders. There are also some truly aquatic spiders, but these are not illustrated. Figs. 7A and B are Cladocera (*Daphnia* group) and Ostracoda (seed shrimps), respectively. These are extremely common and abundant under certain conditions; they range in size from a fraction of to several mm in length. A fairy shrimp is illustrated in Fig. 7C. They often occur in seasonal ponds, as do the tadpole shrimp in Fig. 7D. Both are in the 2-3 cm size range. An aquatic pill bug (Isopoda) is shown in Fig. 7E. Copepods, with characteristic egg sacs, forked "tail" and single eye (*Cyclops* of mythology) are common and range in size from 1-5 mm. Mites (Fig. 7H) range from minute to 4 or 5 mm in diameter. Amphipods (Scuds, Fig. 7G) are often a cm or more in length. Crayfish (Decapoda) are often several inches in length. For coverage of these groups see Pennak (1953) or Edmundson (1959) in the reference section.

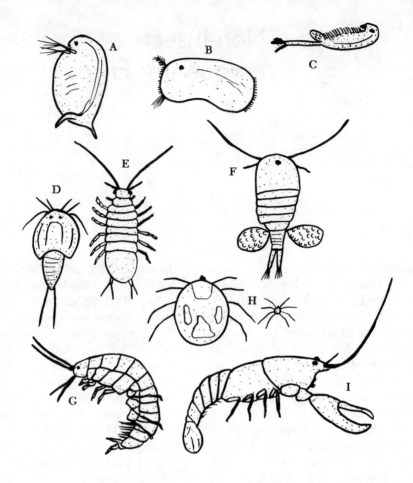

non-insects

Figure 7   A. Water-Flea (Cladocera); B. Ostracoda; C. Fairy Shrimp; D. Tadpole Shrimp; E. Isopod; F. Copepod; G. Amphipod; H. Water mites; I. Crayfish.

Morphology of Insects

A knowledge of basic insect morphology is essential for the use of keys. An insect is divided into three regions, the *head, thorax,* and *abdomen* (Fig. 8). The head bears *antennae, compound eyes,* the *ocelli* or simple eyes, and the mouthparts (Fig. 10). The thorax is made up of three segments indicated by the prefixes *pro-, meso-,* and *meta-,* from anterior to posterior. The thorax is essentially square in cross section (Fig. 9). The top is called the *notum,* the sides the *pleura,* and the bottom the *sternum.* Words compounded from these terms are used to indicate regions of the thorax. For example, the *prosternum* is the ventral side of

the anterior segment of the thorax and the *mesonotum* is the dorsal side of the middle segment. The term *metathorax* refers to the entire posterior segment of the thorax.

The thorax bears the appendages of locomotion. The thoracic legs, from point of attachment to the tip, consist of the *coxa, trochanter, femur, tibia, tarsus,* and *tarsal claws* (Figs. 8, 216). The wings usually consist of thickened veins with membranous areas between. The longitudinal veins conform to a pattern and have been given names and standard letters for abbreviation (Fig. 11). The *costa* (C) forms the front edge of the wing. The *sub-costa* (Sc) follows and may be branched at the tip. The *radius* (R) and *radial sector* (Rs) may have up to 5 branches (R_1-R_5). The *media* (M) primitively branches twice, potentially giving rise to M_1-M_4. The

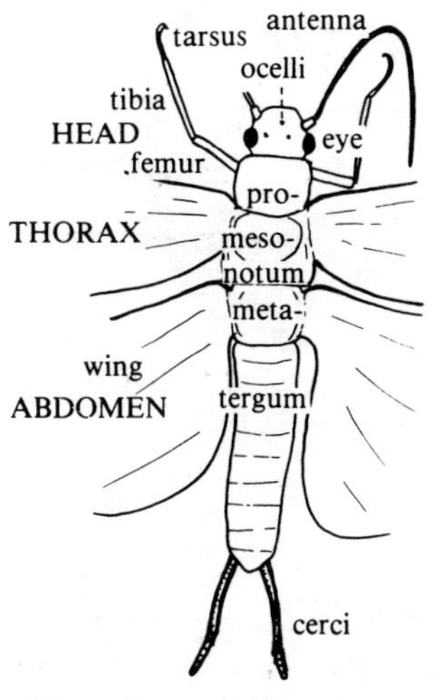

Figure 8   Major features of an insect, diagrammatic.

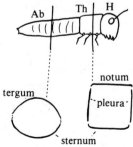

Figure 9   Insect thorax (TH) and abdomen (Ab), diagrammatic.

cubitus (Cu) may have up to three branches, and there may be up to 4 *anal* veins. Major crossveins are named by using the letters of the veins they connect. For example, m-cu connects the media and cubitus. As can be seen in Fig. 11B, most wings are highly modified from the basic plan. Diagrams are presented in the key where needed.

The abdomen is basically a series of ring-like segments which lack appendages but which usually bears *cerci* and *genitalia* at the apex. Many larvae have prolegs or false legs on the abdomen. In cross section (Fig. 9) the abdomen consists of a dorsal *tergum* and the ventral *sternum*. The pleural area, where the tergum and sternum join, may be membranous or completely fused and impossible to distinguish.

The regions of the head and the mouthparts are shown in Fig. 10. The *labrum* is a roof-like flap over the top of the mouth. Just behind the labrum are the chewing *mandibles*. The paired *maxillae* with a complex structure and usually with long palps are behind the mandibles. They are used to manipulate food. The *labium* is essentially a pair of fused maxillae that forms the back or bottom of the mouth.

All of the structures discussed above are usually modified. An example is the Psephenid larva (Coleoptera) in Fig. 12. On close examination, an interpretation of structures is usually possible.

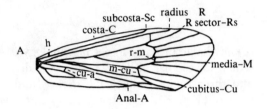

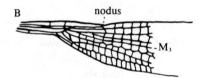

Figure 11 Wings and veins (A. Trichoptera; B. Odonata); see text.

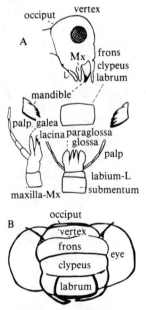

Figure 10 A. Head and mouthparts, diagrammatic; B. front view of face of dragonfly.

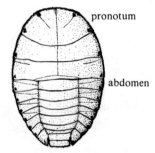

Figure 12 Psephenidae (Coleoptera), showing modifications in basic insect plan.

Collecting and Preserving Aquatic Insects

Many aquatic insects can be collected by using no more than the bare hands. Much can be found by simply turning over stones or examining aquatic vegetation. Around springs, seeps and in small streams this is a very effective method. The variety and quality of the catch however will increase considerably with the addition of a few basic items which are usually readily available. Fig. 13 shows the basic collecting equipment which I prefer. Fishermen's chest waders (A) and a long handled net (H) will allow access to all but deep water habitats; the waders are especially useful when the water is cold or the substrate rough and sharp. Canvas or similar shoes can also be worn to protect the feet.

The net can be passed through beds of reeds or swept through the open water. In streams or rivers the net can be held stationary while the upstream substrate is raked and stirred with the foot, the insects being carried into the net by the current. In ponds or on lake shores a current of water can be created with the foot, thus carrying aquatic insects into the net. A lighter net with a deeper bag is useful for capturing terrestrial stages or shore inhabitants. Both are available in biological supply catalogs (see Appendix).

It is useful to have a large bag (B) of some type, such as those used by fishermen or hikers, for carrying the smaller collecting equipment. Some collectors prefer a standard fisherman's box for the same purpose. The bag or box can contain a variety of bottles and vials of 80% ethyl alcohol, rubbing alcohol, or dilute formalin (13 C). Eighty per cent alcohol is

Figure 13   Collecting equipment, see text.

best (it will soon be diluted to 70% or less by the wet insects collected). Once a sample has been taken in the large net, it can be gently washed to remove excess mud by moving net back and forth in the water. The contents can then be taken to a shallow white enamel or plastic pan (J) (a useful type is available from photo developing supply stores) which has half an inch or an inch of clear water in the bottom, and the contents of the net can be dumped into the pan. The organisms will begin to untangle themselves from the plant and algal material, swimming freely and visibly against the white background. One can get a general impression of what is present. Either the entire collection can be saved (drain through the net and place the collection in a large collecting jar) or individual specimens can be sucked out with a "turkey baster" (D) and squirted gently into a small fine mesh aquarium net (F). The specimens can then be transferred to a small vial with forceps (E) or with the fingers. A "lifter" (K) is also very useful. It is simply a piece of fine wire cloth which has been folded into a spoon-like shape. It can be used to pick out selected specimens. Two good habits are the notebook (I) and the thermometer (G). It adds interest to collecting if notes are kept on each sample site. It is essential to record the exact location and date. It is desirable to note the type of habitat, unusual features, weather, temperature, and activity of adults. In general, the more that is recorded the better, depending on the interests of the collector.

One item not shown in Fig. 13 is the killing jar which is used for immobilizing adults. Any wide mouth jar can be made into a killing jar. Absorbent material, ranging from paper towels to plaster of paris, is placed in the bottom of the jar, and a small amount of ethyl acetate or chloroform is poured into the absorbent material. Collected specimens are placed in the jar and the lid quickly closed. The exact model to be used depends on the needs and preferences of the user. My choice is an easily carried 1" x 8" test tube with a cork stopper. Masking tape is wound around the bottom and the neck to reduce risk of breakage.

SPECIAL COLLECTING TECHNIQUES AND EQUIPMENT

A number of other collecting methods have been devised. The *handscreen* is a simple and effective device consisting of a piece of window screen about 2 feet square. Handles about a foot longer than the screen are stapled or nailed to two opposite sides. The bottom of the handles is thrust into the substrate and the top of the handles are held with one hand while the foot is used to "kick" a stream of water and disturbed substrate through the mesh of the net. This is a very effective but inexpensive and easily made device.

A *drift net,* in principle, is a deep-bagged net of any type which is anchored, mouth upstream, and allowed to remain several hours or overnight while current flows through. This method can yield huge collections containing a high percentage of the members of a stream community.

An *artificial substrate* involves bundling up a collection of brick fragments, soaked wood, and other suitable material into a manageable size, perhaps using poultry netting to hold it together. It is placed on a river or lake bottom for a number of days until it is colonized by the surrounding insect community. The sampler substrate is then retrieved with a rope or wire previously attached. Rare and unusual organisms are sometimes collected by this means.

Various dredges and samplers are commercially available (See section on suppliers). Some are lowered to lake or river bottoms where metal jaws, triggered by a weighted messenger, enclose a sample of the substrate and organisms in a metal box. Others consist of a metal frame which encompasses a section of stream or river bottom in shallow water to give a quantitative sample. It is also possible to devise equipment. A short length of stovepipe can be used to isolate a portion of pond or stream bottom. Organisms are dipped out with a net.

The sample is quantitative and the exact type of substrate can be recorded.

STORAGE AND SPECIAL PREPARATION

After the specimens have been collected, they must be sorted and prepared for study (Fig. 14). The alcohol should be changed to fresh 70% since dilutions probably occurred during collecting. Almost everything in an aquatic collection can be stored in 70% alcohol (higher concentrations make the specimens brittle). Many biologists add a small amount of glycerine to insure that the specimen can be salvaged if it should accidently dry out. Exceptions to storage in alcohol are that adult Odonata and Lepidoptera should be collected in a killing jar and mounted on pins in the standard way (D). (It is possible to mount and store adults of all groups dry, on pins, if desired.) Other exceptions are that the larvae of Lepidoptera, Megaloptera, (especially Corydalidae) and many Diptera become blackened and shrivelled if placed in alcohol directly. Better specimens can be obtained by keeping them alive until it is possible to drop them for a minute into boiling water. Later they are stored in alcohol. The overall objection to storage in alcohol is that it causes color to fade. In general, however, there is no alternative.

All vials, pinned specimens, and slides must carry the basic data: locality (state or province, county, proximity to town or city), date, and collector's name. Identifications can be placed on the same or a separate label. Avoid placing a number of small labels in the vial. They are not only difficult to read, but chop up specimens. Inch-square labels curled around the inside of the vial are preferred. Vial racks (E) can be constructed or purchased and the collection arranged according to the wishes of the collector. Another means of storage is to use cotton plugs in place of a cork or neoprene stopper in the vials. Widemouth jars are filled with vials. The large jar can then be filled with alcohol and sealed.

Pinned specimens are best placed in storage boxes of various types (C) either specially purchased, or improvised. Cigar boxes were commonly used when they were more readily available. A piece of cork is glued to the bottom to receive the pins. It is often desirable to make slides (B) of various insect parts. I use nothing but Canada balsam for making slides. If cellosolve (ethylene glycol monoethyl ether) is added to balsam, dissected parts can be introduced directly from 70% alcohol with no need for dehydration through absolute alcohol. Specimens will clear within hours in the balsam-cellosolve. If specimens are very dark they can be cleared in strong KOH (potassium hydroxide) in water (overnight if cold, a few minutes if boiled), then washed and reintroduced to alcohol before mounting in balsam-cellosolve. Wings are best mounted dry, using glued paper strips to hold the cover slip in place (B).

Another special method of preparation allows dry mounting of specimens (especially nymphs) previously preserved in alcohol. Place a specimen that has been preserved in alcohol on paper towels or filter paper that is lightly soaked with acetone. Cover the filter paper and specimen with an inverted jar or

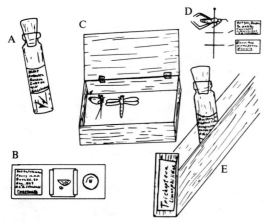

Figure 14 Equipment for storage and preparation, see text.

similar chamber to prevent evaporation loss. After a few hours the alcohol and water in the specimen will be replaced by acetone. Place the specimen in the direct heat of a light bulb. As the specimen dries, vapors from the evaporating acetone maintain the inflation of soft body parts. In a few minutes, the specimen is ready to be glued to a standard insect pin and stored as a dry specimen.

Results vary, but in general, bristles and surface features become prominent, and specimens can be easily oriented to observe difficult-to-see parts. Specimens are not fragile after this treatment and usually keep the same orientation and proportions that they assumed when placed in alcohol.

REARING AQUATIC INSECTS

A major contribution that can be made by any aquatic entomologist is the rearing and association of immature and adult stages. Most species are described on the basis of adults. Probably no more than 25% of the immature stages of aquatic species are associated with their respective adult stages. In other words, most nymphs and larvae cannot be identified to species because they have never been de-

scribed. Associations obtained by rearing individual specimens are essential to the eventual construction of keys to species for nymphs and larvae.

Mature nymphs can be placed in screen cages, either in aerated aquaria or attached to floats in a natural environment. Part of the cage must project above the water to allow the adult to emerge. The emerged adult and cast nymphal skin should be placed in the same vial of alcohol and carefully labelled.

Pupae of caddisflies (and probably other insects) can be successfully reared by taking the pupae from the water and placing it on moss or lightly soaked absorbent material in a small vial. Larval sclerites are usually in the pupal case, along with the pupal skin. Again, all material should be preserved in the same vial and carefully labelled. Specialists usually welcome quality material, and anyone can make a contribution to aquatic entomology. Addresses of specialists can be found on papers they have published (section on Journals and Societies).

The method of rearing is slightly different for each group; there is no absolutely correct way, and experimentation may be required. However, rearing is a major aspect of aquatic entomology.

Pollution and Aquatic Insects as Indicator Organisms

An interest in environmental quality is one of the major reasons for the study of aquatic insects, and a brief discussion seems appropriate here. The idea of using the aquatic insect community to "indicate" the degree of purity or pollution of a body of water is over half a century old. This idea is based on the assumption that natural and undisturbed aquatic systems will have a great diversity of species. The community will also include sensitive species with very little tolerance for disturbance. In contrast, ecologically stressed systems will have comparatively fewer species (but often in large numbers) and sensitive species will be absent. The application of the method requires special knowledge when put in practice. There are situations, for example, where two streams will be entirely different based on insect inhabitants, but both streams may be totally free of unnatural disturbances. In practice, the system may be almost as complicated as a physician diagnosing a disease. Considerable study, background information, and experience are necessary to correctly evaluate the situation.

Some methods of using indicator organisms involve rather complex mathematical analysis, based on information theory, and calculation of a diversity index. Further information and further references can be found in Hilsenhoff (1977), and are beyond the scope of this discussion. The above reference also dis-cusses a method of evaluating aquatic habitats which can be applied by users of this book. This method is as follows. Taxa of aquatic insects are given a numerical score ranging from 0 to 5 (Table 4). The value is based on field observation, and if possible, laboratory experiments. Zero taxa are extremely intolerant of pollution; taxa with a score of 2 through 4 are tolerant; taxa which can survive great amounts of pollution are scored 5. Hence, almost all Plecoptera are 0 and cannot tolerate pollution. Many Chironomidae are 5. Table 4 gives a list of taxa, adapter from Hilsenhoff (1977) and their pollution tolerance score.

In many regions it will be necessary to determine the score of local taxa by field study, and these can be added to the list.

To use the method, select an aquatic habitat of interest. The procedure for evaluating the habitat is to collect about 100 organisms from a variety of areas within the habitat. The organisms are identified and the total number (n_i) of each is recorded. Then use the formula for biotic index (BI):

$$BI = \frac{\Sigma \, n_i a_i}{N}$$

where n_i is the number of specimens in each taxonomic group, a_i is the pollution tolerance score (Table 4) for that taxonomic group, and N (usually 100) is the total number of arthro-

Table 4 Biotic Index, scores for individual taxa (see text) (adapted from Hilsenhoff, 1977)

Taxonomic Group	Pollution tolerance Score
Plecoptera	
(Stoneflies)	
Capniidae, Leuctridae, Nemouridae	0
Taeniopterigidae	1
Chloroperlidae, Perlidae, Perlodidae	0
Pteronarcidae	1
Ephemeroptera	
(Mayflies)	
Baetidae	2.5
Baetiscidae	2
Caenidae	
Caenis	4
Brachycerus	2
Ephemerellidae	0.5
Ephemeridae	1.5
Heptageniidae	
Epeorus	0
Rhithrogena	0
Stenonema	2
Heptagenia	2
Leptophlebiidae	
Paraleptophlebia	1
Leptophlebia	3
Polymitarcidae **(Ephoron)**	1
Siphlonuridae	2
Tricorythidae	2
Odonata	
(Dragonflies and Damselflies)	
Agrionidae	0
Cordulegastridae	0
Aeshnidae	1
Gomphidae	0.5
Lestidae	3
Coenagrionidae	3
Trichoptera	
(Caddisflies)	
Brachycentridae	0.5
Helicopsychidae	1
Hydroptilidae	3
Molannidae	1
Philopotamidae	0
Rhyacophilidae	0
Hydropsychidae **(Cheumatopsyche)**	4
Polycentropodidae **(Neuroclipsis)**	4
Megaloptera	2
(Alderflies and Dobsonflies)	
Coleoptera	
(Beetles)	
Elmidae	2
Psephenidae	2
Diptera	
(True Flies)	
Blephariceridae	0
Ceratopogonidae **(Bezzia)**	3
*Chironomidae **(Chironomus)**	5
Empididae	4
Ephydridae	4
Rhagionidae	2
Tabanidae	2
Tipulidae	2
**Syrphidae	5

*Some members of this family have a score of 0.

**(Not from Hilsenhoff; added by DML)

Table 5 Water Quality (see text) (adapted from Hilsenhoff, 1977)

Biotic Index	Water Quality
under 1.75	Excellent, no disturbance
1.75-2.25	Good, possibly some disturbance or organic enrichment
2.25-3	Fair, probably some disturbance
3.0-3.75	Poor, significant disturbance
over 3.75	Very poor, gross disturbance

pods in the sample. Thus, for each taxon, multiply the number of individuals collected by the score in Table 4 for that group. Add the values for all the groups present, and divide by the total number of insects collected. This will give the biotic index (BI). Table 5 gives an evaluation of the water quality based on the biotic index thus calculated. The test should be made several different seasons of the year before final conclusions are drawn. These tables are based on streams and will require modification for lake and pond studies.

How to Use This Book

In most cases a 10X hand lens is adequate for using the keys in this book. In some cases the naked eye is sufficient, but a dissecting microscope is ideal.

The use of a dichotomous key involves comparing a specimen with two contrasting and mutually exclusive statements. The correct statement is selected and numbers and couplets are followed until a name is reached. To begin, turn to the key to Orders and read couplets 1a and 1b. If functional wings are present on the specimen, choose 1a and proceed on to couplet 19. Read couplets 19a and b. If the front wings are hard and shell-like, proceed on to 20, and read 20a and 20b, then 21a and 21b. Order Coleoptera has hard and shell-like front wings. The Order has been identified. Turn to the key to Families for adult Coleoptera, and proceed comparing the contrasting statements as before until a Family name is reached.

TEXT

A brief discussion follows in each couplet after a taxonomic name. The listing of additional characters should help to confirm the identification. To obtain complete information for a taxon, it will be necessary to read both adult and larval or nymphal sections of the key (when both are present). In general, the adult section contains information on number of species, distribution, evolutionary relationships, and additional features which characterize the group. The larval or nymphal section stresses information on ecology and behavior as well as taxonomic characters.

REGION COVERED

An attempt has been made to include all Families (and occasionally genera and species) of aquatic insects found in North America north of Mexico. It is emphasized that in this book, North America means north of Mexico, and many statements about distribution and number of genera and species will be inaccurate if this is not noted.

TAXONOMY AND NAMES

The following is the basic taxonomic hierarchy which is used for all animals. Standard endings are used for most categories of taxonomic names.

Category	Standard Ending	Example
Class	usually *-a*	Insecta, the insects
Order	*-optera* (in Insects)	Plecoptera, Stoneflies
Superfamily	*-idea*	Dryopoidea, dryopid beetles
Family	*-idae*	Culicidae, mosquitoes
Subfamily	*-inae*	Libellulinae, a group of Dragonflies

Genus and species names do not have standard endings but are either underlined or written in italics. A genus and species name is usually followed by the author of the name, but the practice has not been followed in this general book. It is customary to abbreviate a genus name after the first time it is mentioned in a discussion.

Comparatively few species names are mentioned in this book, either in the text portion or in relation to the figures. With a con-servative estimate of 6000 species of aquatic insects in North America, multiplied by two for immature and adult stages, it is usually misleading to discuss a particular species when the reader almost certainly does not have the same species in hand. Therefore, in any except the smallest groups, both discussion and figures refer to levels above the species level, and in this way statements in the text and figures will be equally appropriate for my material and for that of the reader.

Taxonomic names above the species level are artificial in the sense that different workers may not agree on whether a given group should have Family (or other) status; for example, it is now customary to treat the former subfamilies (Capniinae, Taeniopteriginae, Nemourinae, Leuctrinae) of the old Nemouridae (Plecoptera) as full Families. In the same way, many former stonefly subgenera are now given full generic status. While opinions may be contradictary on questions of this type, taxonomic names are real in the sense that they indicate degree of relationship and carry implications of common behavior, biology and ancestry. The value of a Family or other taxonomic name is the great wealth of information it carries with it. Mention of the name "Culicidae" (mosquitoes) should illustrate the point.

General References

ECOLOGY OF AQUATIC INSECTS

Bay, E. C. 1974. Predator-Prey Relationships Among Aquatic Insects. Ann. Rev. Entomol. 19: 441-453.

Cummins, K. W. 1973. Trophic Relations of Aquatic Insects. Ann. Rev. Entomol. 18: 183-206.

Hilsenhoff, W. L. 1977. Use of Arthropods to Evaluate Water Quality of Streams. Tech. Bull. No. 100. Dept. Nat. Res., Madison, Wisconsin. pp. 1-15.

Hynes, H. B. N. 1970. The Ecology of Running Waters. Univ. of Toronto Press. 555 pp.

Hynes, H. B. N. 1970. The Ecology of Stream Insects. Ann. Rev. Entomol. 15: 25-42.

Lehmkuhl, D. M. 1972. Change in Thermal Regime as a Cause of Reduction of Benthic Fauna Downstream of a Reservoir. J. Fish. Res. Bd. Canada 29: 1329-1332.

Tauber, M. J. and C. A. Tauber. 1976. Insect Seasonality: Diapause Maintenance, Termination, and Postdiapause Development. Ann. Rev. Entomol. 21: 81-107.

Waters, T. F. 1972. The Drift of Stream Insects. Ann. Rev. Entomol. 17: 253-272.

Wiggins, G. B. 1973. A Contribution to the Biology of Caddisflies (Trichoptera) in Temporary Pools. Life Sciences Cont., Royal Ontario Museum, No. 88. 28 pp.

GENERAL TAXONOMY AND BIOLOGY

Bland, Roger. 1978. How to Know the Insects, Wm. C. Brown Co., Dubuque, Iowa. 409 pp.

Borror, D. J., D. M. DeLong, and C. A. Triplehorn. 1976. An Introduction to the Study of Insects. 4th ed. Holt, Rinehart, and Winston, 852 pp.

Borror, D. J. and R. E. White. 1970. A field guide to the insects of America north of Mexico. Houghton Mifflin, Boston, 404 pp.

Cheng, Lanna (ed.). 1976. Marine Insects. American Elsevier Publ. Co., New York. 581 pp.

Chu, H. F. 1949. How to Know the Immature Insects. Wm. C. Brown Co., Dubuque, Iowa. 234 pp.

Edmondson, W. T. (ed.). 1959. Freshwater Biology, 2nd Ed. Wiley, New York. 1248 pp.

Merritt, R. W. and K. W. Cummins, eds. 1978. An introduction to the aquatic insects of North America. Kendall/Hunt, Dubuque, Iowa. 441 pp.

Pennak, R. W. 1953. Freshwater Invertebrates of the United States. Ronald Press, New York. 769 pp.

Peterson, A. 1951. Larvae of insects, Pt. II. Coleoptera, Diptera, Neuroptera, Siphonaptera, Mecoptera, Trichoptera. Edwards Bros., Ann Arbor, Mich. 416 pp.

Usinger, R. L. (ed.). 1956. Aquatic Insects of California. Univ. of Calif. Press, Berkeley. 508 pp.

METHODS

Beirne, Bryan P. 1955. Collecting, Preparing, and Preserving Insects. Canada Dept. Ag., Ottawa. 133 pp.

Knudsen, J. W. 1972. Collecting and Preserving Plants and Animals. Harper and Row, New York. 320 pp.

Mayr, Ernst, E. G. Linsley, and R. L. Usinger. 1953. Methods and Principles of Systematic Zoology. McGraw-Hill, New York. 336 pp.

Needham, J. G. 1937 (reprint, 1959). Culture methods for invertebrate animals. Dover Publications, New York. 571 pp.

Smith, R. C. and R. H. Painter. 1966. Guide to the Literature of the Zoological Sciences, 7th ed. Burgess Publ. Co., Minneapolis. 238 pp.

COLLEMBOLA

Christiansen, K. 1964. Bionomics of Collembola. Ann. Rev. Entom. 9: 147-178.

Gisin, Hermann. 1960. Collembolenfauna Europas. Museum D'Histoire Naturelle. Geneve. 312 pp.

Maynard, E. A. 1951. A Monograph of the Collembola of New York State. Comstock Publ. Co., Inc., Ithaca, N. Y. 339 pp.

Scott, D. B. 1956. Aquatic Collembola. pp. 74-78, in Aquatic Insects of Calif. R. L. Usinger, ed., Univ. of Calif. Press.

EPHEMEROPTERA

Allen, R. K. 1973. Generic revisions of Mayfly Nymphs. 1. *Traverella* in North and Central America (Leptophlebiidae). Ann. Ent. Soc. Amer. 66: 1287-1295.

Allen, R. K. and George F. Edmunds, Jr. 1965. A revision of the genus *Ephemerella* (Ephemeroptera: Ephemerellidae) VIII. The subgenus *Ephemerella* in North America. Misc. Publ. Ent. Soc. Amer. Vol. 4, No. 6, pp. 243-282.

Berner, Lewis. 1950. The Mayflies of Florida. Univ. of Calif. Press, Univ. of Florida Studies Vol. IV, No. 4, pp. 1-267.

Burks, B. D. 1953. The Mayflies, or Ephemeroptera, of Illinois. Bull. Ill. Nat. Hist. Survey 26: 1: 10216.

Edmunds, George F., Jr. 1957. *Metretopus borealis* (Eaton) in Canada (Ephemeroptera: Ametropodidae). Can. J. Zool. 35: 161-162.

Edmunds, George F., Jr. 1972. Biogeography and Evolution of Ephemeroptera. Ann. Rev. Entomol. 17: 21-42.

Edmunds, George F., Jr., R. K. Allen, and W. L. Peters. 1963. An annotated key to the nymphs of the families and subfamilies of mayflies (Ephemeroptera). Univ. of Utah Biol. Service Vol. XIII, No. 1, Salt Lake City. 49 pp.

Edmunds, George F., Jr., Lewis Berner, and J. R. Traver. 1958. North American Mayflies of the Family Oligoneuriidae. Ann. Ent. Soc. Amer. 51: 375-382.

Edmunds, George F., Jr. and R. W. Koss. 1972. A review of the Acanthametropodinae with a description of a new genus (Ephemeroptera: Siphlonuridae). Pan-Pac. Entom. 48: 136-144.

Edmunds, George F., Jr., S. L. Jensen and Lewis Berner. 1976. The mayflies of North and Central America. U. of Minnesota Press, Minneapolis. 330 pp.

Edmunds, G. F., Jr. and J. R. Traver. 1959. The classification of the Ephemeroptera, I. Ephemeroidea: Behningiidae. Ann. Ent. Soc. Amer. 52: 43-51.

Koss, R. W. 1968. Morphology and taxonomic use of Ephemeroptera eggs. Annals Ent. Soc. Amer. 61: 696-721.

Lehmkuhl, D. M. 1973. A new species of *Baetis* (Ephemeroptera) from ponds in the Canadian arctic, with biological notes. Can. Ent. 105: 343-346.

Lehmkuhl, D. M. 1976. Mayflies (list of Saskatchewan species). Blue Jay 34: 70-81.

Lehmkuhl, D. M. 1976. Additions to the taxonomy, zoogeography and biology of *Analetris eximia* (Ephemeroptera). Can. Ent. 108: 199-207.

McCafferty, W. P. 1975. The burrowing mayflies (Ephemeroptera: Ephemeroidea) of the United States. Trans. Amer. Ent. Soc. 101: 447-504.

Needham, J. G., J. R. Traver and Yin-Chi Hsu. 1935. The biology of mayflies. Comstock Publ. Co., Ithaca, N. Y., 759 p.

Traver, Jay R. and G. F. Edmunds, Jr. 1968. A revision of Baetidae with spatulate-clawed nymphs (Ephemeroptera). Pacific Insects 10(3-4): 629-677.

ODONATA

Johnson, C. 1972. The damselflies (Zygoptera) of Texas. Bull. Fla. State Mus. Biol. Sci., 16(2): 55-128.

Needham, J. G. and H. B. Heywood. 1929. A handbook of the dragonflies of North America. Charles C. Thomas, Springfield, Illinois. 378 pp.

Needham, J. G. and M. J. Westfall, Jr. 1955. A manual of the dragonflies of North America. Univ. Calif. Press. 615 pp.

Paulson, D. R. and R. W. Garrison. 1977. A list and new distributional records of Pacific Coast Odonata. Pan-Pac. Ent. 53: 147-160.

Walker, E. M. 1953 The Odonata of Canada and Alaska. The Zygoptera-Damselflies. Vol. I, Univ. Toronto Press. 292 pp.

Walker, E. M. 1958. The Odonata of Canada and Alaska. The Anisoptera—Four Families. Vol. II, Univ. Toronto Press. 318 pp.

Walker, E. M. and P. S. Corbet. 1975. The Odonata of Canada and Alaska, Libelluli-dae. Vol. III, Univ. Toronto Press. 307 pp.

ORTHOPTERA

Brooks, A. R. 1958. Acridoidea of southern Alberta, Saskatchewan, and Manitoba (Orthoptera). Canad. Ent. Suppl. 9. pp. 1-92.

Hebard, Morgan. 1934. The Dermaptera and Orthoptera of Illinois. Bull. Ill. Nat. Hist. Surv. Vol. 20, Art 3, pp. 125-279.

Helfer, J. R. 1953. The Grasshoppers, Cockroaches, and their Allies. Wm. C. Brown Co., Dubuque, Iowa. 350 pp.

PLECOPTERA

Claassen, P. W. 1931. Plecoptera Nymphs of America (North of Mexico). Thomas Say Foundation, Baltimore, 199 pp.

Frison, T. H. 1935. The Stoneflies, or Plecoptera, or Illinois. Bull. Illinois Nat. Hist. Surv. 20: 281-471.

Gaufin, A. R., W. E. Ricker, M. Miner, P. Milam, and R. A. Hays. 1972. The Stoneflies (Plecoptera) of Montana. Trans. Amer. Ent. Soc. 98: 1-161.

Harden, P. H. and C. E. Mickel. 1952. The Stoneflies of Minnesota (Plecoptera). U. of Minnesota. Ag. Expt. Station Bull. 201. 84 pp.

Hitchcock, S. W. 1974. The Plecoptera, or Stoneflies of Connecticut. Pt. VII, Insects of Conn. Bull. 107, State Geol. and Nat. Hist. Surv., Hartford. 262 pp.

Illies, J. 1965. Phylogeny and Zoogeography of the Plecoptera. Ann. Rev. Entom. 10: 117-140.

Jewett, S. G., Jr. 1959. The stoneflies (Plecoptera) of the pacific northwest. Oregon State Univ. Press, Studies in Entomology, No. 3. 95 pp.

Lehmkuhl, D. M. 1971. Stoneflies (Plecoptera: Nemouridae) from temporary Lentic habitats in Oregon. Amer. Midl. Nat. 85: 514-515.

Zwick, P. 1973. Insecta: Plecoptera, Phylogenetisches System und Katalog. Das Tierreich. Leiferung 94. Walter de Gruyter and Co., Berlin. 465 pp.

HEMIPTERA

Brooks, A. R. and L. A. Kelton. 1967. Aquatic and Semiaquatic Heteroptera of Alberta, Saskatchewan, and Manitoba (Hemiptera). Memoirs Ent. Soc. Canada No. 51. 92 pp.

Cheng, Lanna. 1977. The elusive Sea Bug *Hermatobates* (Heteroptera). Pan-Pac. Entom. 53: 87-97.

Froeschner, R. C. 1961. Contributions to a synopsis of the Hemiptera of Missouri, Part V (Aquatic families). Amer. Midl. Nat. 67: 208-240.

Gittelman, S. H. 1977. Leg segment proportions, predatory strategy and growth in backswimmers (Hemiptera: Pleidae, Notonectidae). J. Kans. Ent. Soc. 50: 161-171.

Hungerford, H. B. 1948. The Corixidae of the Western Hemisphere (Hemiptera). Univ. Kan. Sci. Bull. 32: 1-827.

Southwood, T. R. E. and D. Leston. 1959. The land and water bugs of the British Isles. Warne and Co., London and New York. 436 pp.

Usinger, R. L. 1956. Aquatic Hemiptera (see Usinger 1956, general literature).

Voigt, W. G. and R. Garcia. 1977. Keys to the *Notonecta* nymphs of the west coast United States Pan-Pac. Entom. 52: 172-176.

MEGALOPTERA

Azam, K. M. and N. H. Anderson. 1969. Life history and habits of *Sialis rotunda* and *S. californica* in western Oregon. Annals Ent. Soc. Amer. 62: 549-558.

Chandler, H. P. 1956. Megaloptera. In: Aquatic insects of California. Ed. by R. L. Usinger. Univ. Calif. Press, Berkeley. P. 229-233.

Evans, E. D. 1972. A study of the Megaloptera of the Pacific Coastal region of the United States. Ph. D. Thesis, Oregon State University, Corvallis. 210 pp.

Flint, O. S. 1964. New species and new state records of *Sialis* (Neuroptera: Sialidae). Ent. News 85: 9-13.

Flint, O. S. 1965. The genus *Neohermes* (Megaloptera: Corydalidae). Psyche 72: 255-263.

Ross, H. H. 1937. Nearctic alderflies of the genus *Sialis* (Megaloptera, Sialidae). Bull. Ill. Nat. Hist. Surv. 21: 57-78.

Townsend, L. H. 1939. A new species of *Sialis* (Megaloptera: Sialidae) from Kentucky. Proc. Ent. Soc. Wash. 4: 224-226.

NEUROPTERA, SISYRIDAE

Brown, H. P. 1952. The life history of *Climacia areolaris* (Hagen), a neuropterous "parasite" of freshwater sponges. Amer. Midl. Nat. 47: 130-160.

Parfin, Sophy I., and A. B. Gurney. 1956. The Spongilla–Flies, with special reference to those of the western hemisphere (Sisyridae, Neuroptera). Proc. U.S. Nat. Mus. 105: 421-529.

TRICHOPTERA

Betten, C. 1934. The Caddis Flies or Trichoptera of New York State. New York State Museum Bull. No. 292, Albany. 576 pp.

Fischer, F. G. J. 1960-1973. Trichopterum Catalogus. I-XV. Nederlandse Vereeniging, Amsterdam.

Flint, O. S. 1960. Taxonomy and Biology of Nearctic Limnephilid larvae (Trichoptera) with special reference to species in eastern U. S. Ent. Amer. No. 40, pp. 1-120.

Lehmkuhl, D. M. 1970. A North American trichopteran larva which feeds on freshwater sponges (Trichoptera: Leptoceridae: Porifera: Spongillidae). Amer. Midl. Nat. 84: 278-280.

Malicky, H. (ed.). 1976. Proceedings of the first international symposium on Trichoptera. Junk Publ., The Hague. 213 pp.

Nimmo, A. P. 1971. The Adult Rhyacophilidae and Limnephilidae (Trichoptera) of Alberta and Eastern British Columbia and their postglacial origin. Quaest. Ent. 7: 3-234.

Ross, H. H. 1944. The Caddis Flies, or Trichoptera, of Illinois. Bull. Ill. Nat. Hist. Surv., Urbana, Vol. 23, Art. 1. 326 pp.

Ross, H. H. 1956. Evolution and Classification of the Mountain Caddisflies. U. of Illinois Press, Urbana. 213 pp.

Ross, H. H. 1967. The evolution and past dispersal of the Trichoptera. Ann. Rev. Entomol. 12: 169-206.

Wiggins, Glenn B. 1954. The Caddis Fly Genus *Baraea* in N. A. (Trichoptera). Cont. Roy. Ont. Mus. No. 39. pp. 1-13 + plate.

Wiggins, Glenn B. 1977. Larvae of North American Caddisfly Genera (Trichoptera). U. of Toronto Press. 410 pp.

LEPIDOPTERA

Lange, W. H., Jr. 1956. Aquatic Lepidoptera, pp. 271-288, i n Usinger, R. L., Ed. Univ. of Calif. Press, Berkeley.

Monroe, E. 1972-3. In Dominick, R. B., *et al.* The Moths of America north of Mexico, Fasc. 13-1, parts A, B, C, Pyraloidea (Pyralidae). E. W. Classey Ltd. and R. B. D. Publication, Inc., London.

HYMENOPTERA

Askew, R. R. 1971. Parasitic Insects. American Elsevier Publ. Co., New York. 316 pp.

Hagen, K. S. 1956. Aquatic Hymenoptera, pp. 289-292, *in* Aquatic Insects of Calif. R. L. Usinger, ed., U. of Calif. Press.

COLEOPTERA

Arnett, Ross H. 1960. The Beetles of the United States. The Catholic Univ. of America Press, Washington, D.C. pp. 1-1112.

Boving, A. G. and F. C. Craighead. 1930-31. Illustrated synopsis of the principal larval forms of the order Coleoptera. Ent. Amer. *11*: 1-351. Reprint Edition 1953.

Crowson, R. A. 1967 (reprint). The Natural Classification of Coleoptera. E. W. Classey, Ltd., Middlesex, England. 214 pp.

Edwards, J. Gordon. 1949. Coleoptera or Beetles east of the Great Plains. Edwards Brothers, Inc. Ann Arbor, Mich. 181 pp.

Gordon, R. D. and R. L. Post. 1965. North Dakota Water Beetles. North Dakota Insects. Publ. No. 5. North Dakota State Univ. Fargo. 53 pp.

Hatch, M. H. 1953, 1957, 1961, 1965, 1971. Beetles of the Pacific Northwest. Parts 1-5. Univ. of Washington Press.

Lindroth, Carl H. 1961-69. The Ground-Beetles (Carabidae) of Canada and Alaska. Opuscula Entomologica, Supplements, in 6 parts. Lund, Sweden.

Matta, J. F. 1974, 1976, 1977. The Insects of Virginia No. 8—Hydrophilidae; No. 10—Haliplidae; No. 12 (with A. G. Michael) Dytiscidae. State University, Blacksburg.

Young, F. N. 1954. The Water Beetles of Florida. U. of Florida Press, U. of Florida Studies,

Biological Science Series, Gainesville, Vol. 5, No. 1. 238 pp.

DIPTERA

Alexander, C. P. 1920. The Craneflies of New York. Pt. II. Biology and Phylogeny. New York (Cornell) Ag. Expt. St. Memoir 38: 691-1133.

Alexander, C. P. 1967. The Craneflies of California. Bull. Cal. Insert. Surv., Vol. 8. U. of Calif. Press. 269 pp.

Carpenter, S. J. and W. J. LaCasse. 1955. Mosquitoes of North America (north of Mexico). U. of Calif. Press, Berkeley. 360 pp.

Cole, Frank R. 1969. The Flies of Western North America. U. of Calif. Press, Berkeley. 692 pp.

Cook, Edwin F. 1956. The Nearctic Chaoborinae (Diptera: Culicidae). U. of Minn. Ag. Expt. Station Bull. 218. 102 pp.

Curran, C. H. 1934. The Families and Genera of North American Diptera. New York. 512 pp.

Exner, K. and D. A. Craig. 1976. Larvae of Alberta Tanyderidae (Diptera: Nematocera). Quaest. Ent. 12: 219-237.

Hamilton, A. L., O. A. Saether, and D. R. Oliver. 1969. A classification of the nearctic Chironomidae. Fisheries and marine service technical Report No. 124, Environment Canada, Freshwater Institute, Winnipeg, Manitoba. 42 pp.

Hogue, C. L. 1973. The Net-winged Midges or Blephariceridae of California. Bull. Calif. Insert Survey, Vol. 15, U. of Calif. Press, Berkeley. 83 pp.

Johannsen, O. A. and L. C. Thomsen. 1934-1937. Aquatic Diptera. Parts I-V. New York (Cornell) Ag. Expt. Station Memoirs 164, 197, 205 and 210.

Mason, W. T. 1968. An Introduction to the Identification of Chironomid Larvae. Federal Water Pollution Control Administration, Cincinnati, Ohio. 89 pp.

Oldroyd, H. 1977. The Suborders of Diptera. Proc. Ent. Soc. Wash. 79: 3-10.

Pechuman, L.L. 1973. Horse and Deer Flies of Virginia. Insects of Virginia No. 6, State University, Blacksburg. 92 pp.

Peters, M. T. and E. F. Cook. 1966. The Nearctic Dixidiae (Diptera). Misc. Publ. Ent. Soc. Amer. 5(5): 233-278.

Peterson, B. V. 1970. The *Prosimulium* of Canada and Alaska (Diptera, Simuliidae). Memoirs Ent. Soc. Canada No. 69. Ottawa. 215 pp.

Stone, Alan. 1965. Guide to the insects of Connecticut. Pt. VI Diptera, 9th Fascicle. Simuliidae and Thaumaleidae. Geol. and Nat. Hist. Survey of Conn. Bull. 97, pt. 4, Fasc. 9. 129 pp.

Stone, Alan, C. W. Sabrosky, *et al.* 1965. A Catalog of the Diptera of America north of Mexico. U. S. Gov't. Printing Office, Washington, D. C. Agr. Handbook 276. 1696 pp.

Key to Aquatic Insect Orders

1a Functional wings present and well developed (adult winged insects) (Fig. 15). 19

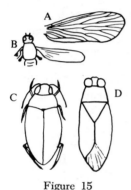

Figure 15

Figure 15 Wings of adult insects. A. Trichoptera; B. Diptera; C. Coleoptera; D. Hemiptera

In some adults the front wings are modified into leathery or hard and shell-like structures which cover the hind wings (Fig. 15 C, D). Careful examination may be necessary to determine if functional wings are present. Among aquatic insects short winged or wingless (brachypterous or apterous) adults are known in Orthoptera, Hemiptera, and Plecoptera. They usually key to the correct Order regardless of the couplet selected here.

1b Wings absent or represented by wing buds only; insect not capable of flight (Figs. 16, 19, 21, 22, 23) (nymphs, larvae, pupae, Apterygota, wingless adults). 2

2a Small insects, usually 1-2 mm, with a forked spring apparatus under abdomen; if spring is absent, then body shape similar to Fig. 16C (Springtails, p. 45). .... **Order Collembola**

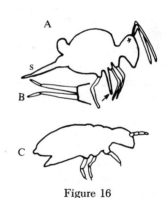

Figure 16

Figure 16 Collembola. A. Symphypleona; B. detail of spring (s); C. Arthropleona

Only this Order among the Apterygota (primitive wingless insects) has aquatic representatives. Most Collembola have a forked structure under the abdomen which is used for jumping. Those which lack the spring are recognized by the characteristic body shape, by the collophore or eversible tube on the venter of the first abdominal segment (arrow, Fig. 16A), and by the presence of an antenna with 6 or fewer segments.

2b Larger, appearance not as in Fig. 16; never with a forked spring-like apparatus under the abdomen. 3

3a   Nymphs; flap-like wing buds present on the thorax (w) (Figs. 18, 19, 21). 4

Young grasshoppers are typical nymphs and the image may be useful in deciding which choice to take. In nymphs tarsi are usually 2-5 segmented, and the thorax is usually distinct from the abdomen, compound eyes are usually present and antenna are usually long and conspicuous. Very young nymphs may lack wing buds and caution is necessary in such cases. Note also that Orthoptera, Hemiptera, and Plecoptera sometimes have wingless adults. In such cases, see Couplets 4a, 7a, and 27a.

3b   Larvae and pupae; external flap-like wing buds either completely absent from the thorax (Figs. 23, 25, 26, 27; larvae) or if present, body is mummy-like, with legs, wings and antennae fused or pressed against the body wall (Fig. 22; pupae). 8

In larvae, the thorax is not clearly differentiated from the abdomen; tarsi, if present, are usually one-segmented. Compound eyes are never present in larva and antennae are almost always short and inconspicuous (exceptions include Helodidae, Coleoptera; Deuterophlebiidae, Diptera; Sisyridae, Neuroptera).

4a   Mouthparts in the form of a segmented beak (Fig. 17) (True Bugs, p. 93). .... Order Hemiptera

Figure 17   Hemiptera, various beaks (A-C); various body shapes (D-J)

Adults and immatures of this Order usually share the same habitats. The segmented beak, formed from fused and modified normal mouthparts, readily separates this Order from all other insects. The beak may be folded tightly backward between the front legs, making it difficult to see; in Corixidae (Fig. 17B) the beak is much flattened. Body shape is extremely varied, and length is from a few mm to over 6 cm.

4b   Mouthparts not in the form of a segmented beak. 5

5a   Mouth covered by an elbowed mask-like labium (Fig. 18A) (Damselflies and Dragonflies, p. 74). .... Order Odonata

The mask may be tightly withdrawn against the underside of the head, but will become obvious by probing with a pin. Damselflies (Suborder Zygoptera) have 3 terminal leaflike gills (Fig. 18B, D) while Dragonflies (Suborder Anisoptera) have terminal spines on the abdomen (Fig. 18E). Length is up to 4 cm.

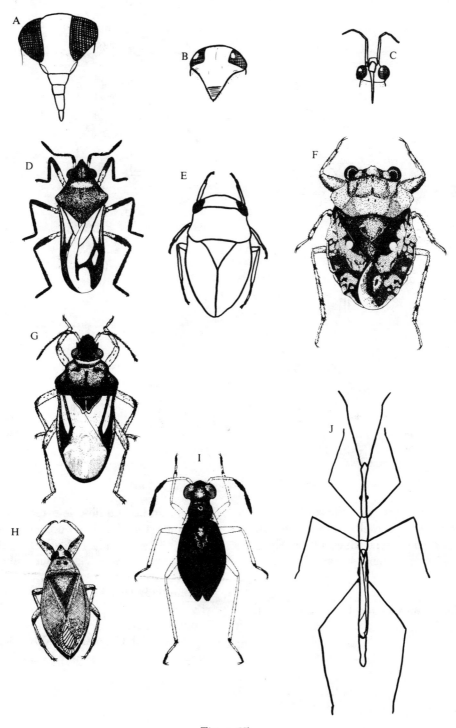

Figure 17

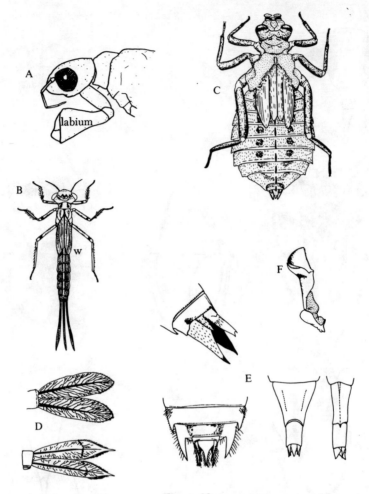

Figure 18

Figure 18 Odonata. A. Mask-like labium; B. Zygoptera (Damselfly); C. Anisoptera (Dragonfly); D. lateral view, Damselfly gills; E. Dragonfly abdomens, 4 examples; F. detail of labium

5b   Mouth not covered by a mask-like labium. 6

6a   Middle and hind legs terminate in a single tarsal claw; abdomen with lateral gills (Fig. 19) (Mayflies, p. 48). .... Order Ephemeroptera

While body shape is extremely varied in mayfly nymphs, the combination of lateral abdominal gills, single tarsal claws, and two or three caudal filaments readily identifies this order. Body length is up to 4 cm.

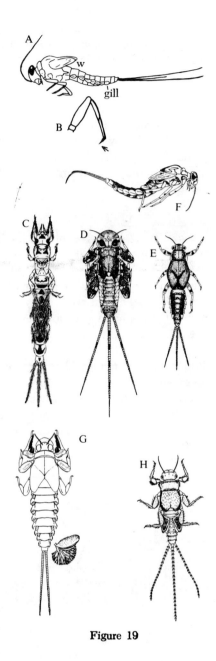

Figure 19

Figure 19 Ephemeroptera. A. Nymph (Baetidae); B. Leg with single tarsal claw (arrow). w—wing bud. C. Ephemeridae; D. Heptageniidae; E. Ephemerellidae; F. Baetidae; G. Oligoneuriidae; H. Tricorythidae.

6b Middle and hind legs with two tarsal claws; abdomen without lateral gills. .. 7

7a Hind legs elongated and modified for jumping (Fig. 20); shore insects (Grasshoppers and Crickets). ....
.... **Order Orthoptera**

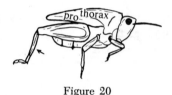

Figure 20

Figure 20 Orthoptera (Tetrigidae) with jumping hind leg (arrow)

The conspicuous jumping hind legs and chewing mouthparts readily separate this Order from all others. The grasses and weeds along the borders of aquatic habitats sometimes abound in various types of grasshoppers, katydids, and crickets, but no North American species can be considered truly aquatic in the way some tropical species are. *Truxalis,* a slant faced grasshopper with flattened antennae, (Family Acrididae), is a reed and cattail inhabiting genus. Two other families are closely enough associated with sand beaches, lake beds, and shores to be included in an aquatic key. Pygmy grasshoppers or grouse locusts

(Tetrigidae) are characterized by the pronotum which extends posteriorly almost to the tip of the abdomen (Fig. 20). They overwinter as adults and can be found in early spring which is unusual among grasshoppers. While not truly aquatic they are frequently collected on the mud flats exposed by receding water of marshes or ponds. Tridactylidae, pygmy mole crickets, burrow in sand along the shores of streams and lakes but are often seen on the surface of the ground. The family contains two species, the 4-5 mm long *Tridactylus minutus* which ranges from New Jersey to Florida and California, and the larger 6-9 mm *T. apicalis* which is known from Ontario, the eastern and western U.S., as well as Mexico and S. America.

7b **Hind legs not elongate and modified for jumping; true aquatic insects (Fig. 21) (Stoneflies, p. 85). ...... Order Plecoptera**

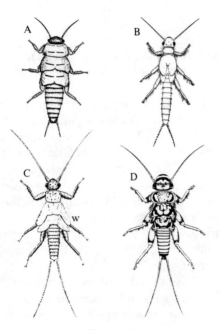

Figure 21

Figure 21 Plecoptera, body shapes; A. Peltoperlidae; B. Chloroperlidae; C. Taeniopterigidae; D. Perlidae

This group is rather uniform in appearance and is easily recognized by the lack of lateral abdominal gills, the two claws on each tarsus, and by the two terminal abdominal filaments. Most are restricted to flowing water. Body length is up to 4 cm.

8a **Larvae; wing buds completely lacking; body often maggot-like, worm-like or caterpillar-like (but see Figs. 30, 32); legs and head present or absent; if legs are present, the tarsi are usually one-segmented; compound eyes are never present. 9**

8b **Pupae; body mummy-like; legs, wings, antennae, and other structures closely appressed to the body wall; compound eyes often visible; sometimes inside a seedlike leathery or silken cocoon (Fig. 22). .... Pupae**

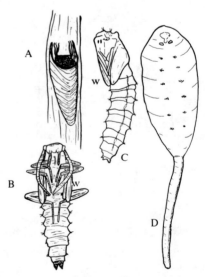

Figure 22

Figure 22 Pupae. A. Simuliidae, Diptera; B. Trichoptera; C. Tabanidae, Diptera; D. Syrphidae, Diptera (w—wing).

The pupa is a "resting" stage in that the insect does not normally move about (there are exceptions such as mosquito "wrigglers"). It is a time of great internal activity and change. While it may eventually be possible to key all pupae to family or even species, keys to pupae are not included here. Often, enough cast-off larval sclerites or adult features are associated with the pupa that the Order or sometimes the Family identification can be made using this key.

9a   **Three pairs of true legs present on the thorax (Figs. 26, 30).** **13**

9b   **Without three pairs of true legs on the thorax; "false legs" may be present on the thorax or abdomen (Figs. 23-25).** ......
.... **10**

True legs are homologous to the 6 legs of adult insects. "False legs," also called pseudopods or prolegs, are structures which resemble legs and may be used for locomotion, but anatomically they are not legs. False legs have developed in various positions in Diptera (Fig. 25). Caterpillars of Lepidoptera (Fig. 26) have 3 pairs of true legs plus several pairs of false legs, and in this case go to 13.

10a   **Headless, legless, white larvae under 2 mm long living as parasites, often inside the eggs or bodies of other insects.** ....
.... **Order Hymenoptera, Wasps**

This order includes the ants, bees, and wasps. Larvae of many wasps are parasitic on the eggs or bodies of other insects. Aquatic insects are included among the hosts. Wasp larvae may

develop inside the host, or they may feed externally (See also Couplet 22a).

(*Note*: Some Diptera are also parasitic; e.g. Tachinid flies have been reared from aquatic moths.)

10b   **Not headless, legless parasites.** **11**

11a   **Without a distinct, sclerotized head capsule (Fig. 23) (True Flies, p. 138).** ....
.... **Order Diptera, in part**

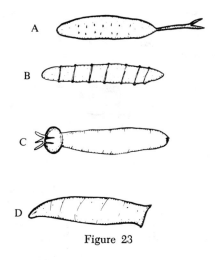

Figure 23

Figure 23 Diptera; body shapes. A. Ephydridae; B. Tabanidae; C. Tipulidae; D. Muscidae

"Higher Diptera" (types resembling the housefly) have maggot-like larvae lacking a head capsule. The larvae of primitive Diptera, such as mosquitoes and midges, have a distinct head capsule. They are highly varied in body shape.

11b   **Larva with a distinct, sclerotized head capsule.** **12**

12a Body oval or elongate, no complex structures at posterior end (Figs. 24, 32B). ....
.... Order Coleoptera, in part,
Family Curculionidae, weevils and Ge-
nus *Cercyon*, Hydrophilidae.

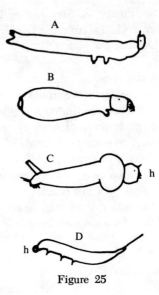

Figure 25

Figure 25 Diptera, body shapes. A. Dixidae;
B. Simuliidae; C. Culicidae; D. Ptychopteridae

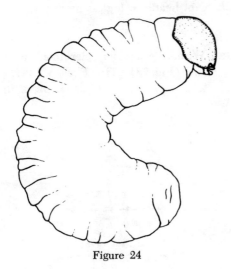

Figure 24

Figure 24 Curculionidae, Coleoptera

Cercyon, a legless group of Hydrophilidae,
has anteriorly directed mouthparts as in Fig.
32B, while in Curculionidae the mouthparts
are ventral (Fig. 24). Weevils are the largest
family of Coleoptera. About 100 species feed
on aquatic plants, burrowing into submerged
stems and roots or living between the stem and
leaf sheath, but not actually contacting the
water. The weevil larva illustrated is not an
aquatic species. See 1a, p. 121.

13a Underside of abdomen with false legs
which bear hooklets arranged in circles
or ovals (Fig. 26) (Moths). ....
.... Order Lepidoptera

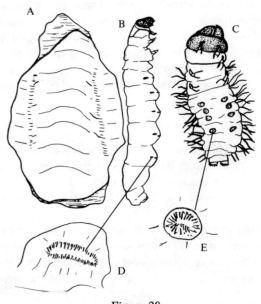

12b Body shape complex, with a breathing
tube or other structures at the end of the
abdomen (Fig. 25) (True Flies, p. 138).
.... Order Diptera, in part

Figure 26

Figure 26   Lepidoptera. A. and B. Leaf case and larva of plant-feeding Nymphulini; C. rock-feeding Argyractini; D. and E. detail of hooklets

A number of families of moths (Cossidae, Cosmopterygidae, Yponomeutidae, Arctiidae, Phalaenidae, Tortricidae, Teneidae, Nepticulidae, Pyralidae, Sphingidae) are known to have some association with water, but it is the Family Pyralidae and the subfamily Nymphulinae (and to a lesser extent Crambinae and Schoenobiinae) that contain the well known aquatic species. The latter 2 subfamilies are stem burrowers in aquatic rushes or grasses. Most Nymphulinae (approximately 14 genera and 43 species) are fully aquatic. The subfamily has been divided into two larval types: the plant feeders (Nymphulini, 10 genera and 23 species) which build cases or tubes of leaves and which have short mandibles and usually no gills (Fig. 26A, B) and the so-called rock feeders (Argyractini, 4 genera and 20 species) which usually live in silk galleries on rocks, have long mandibles, usually have long lateral gills (Fig. 26C) and graze on diatoms or algae in lakes or streams. Although these moths are relatively common and widespread, knowledge of their taxonomy, biology, and distribution is fragmentary. A few species have been well studied; others not at all. Adults can often be easily reared from larvae if mature larvae are placed in moist rearing containers. See also Couplet 23a.

13b   Abdomen without circles of hooklets on underside. **14**

14a   Small larvae, 1/4 inch or less, with mouthparts in form of needle-like stylets; living on freshwater sponges (Fig. 27) (Spongilla Flies). ....
.... **Order Neuroptera, Sisyridae**

Figure 27

Figure 27   Sisyridae, Neuroptera

Larvae of spongilla flies are parasitic on freshwater sponges, especially the sponge genera *Meyenia* and *Spongilla*. Fluid from sponge cells is sucked in by a tube formed from the closely appressed and grooved maxillae and mandibles. The gut is closed and a fecal pellet is not deposited until after adult emergence. Eggs are laid in small silk covered clusters on objects overhanging water. Eggs hatch in about two weeks and the larvae drop to the water. They eventually settle on a sponge and pass through three larval instars. Third instar larvae leave the water, sometimes crawling as much as 50 feet from the water, and spin a cocoon on an appropriate object—tree bark, grass stems, bridges, or other objects. Cocoons are about 3-6 mm in length. Adults may emerge in a week, and there may be three generations a year. These insects apparently overwinter in a pre-pupal stage in a silk case. In mild climates larvae may be present throughout the winter. (See also Couplet 30a).

14b   Mouthparts not in form of needle-like stylets. **15**

15a   Body form very similar to Fig. 30, 7 or 8 abdominal segments with lateral filaments, *but* lacking the single terminal filament, *and* end of abdomen with 4 recurved claws and paired filaments (Fig.

28) (Whirlygig Beetles, p. 122). ....
...... **Order Coleoptera, Family Gyrinidae**

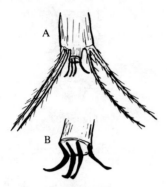

Figure 28

Figure 28 Gyrindae. A. ventral view of last segment of abdomen; B. detail of 4 hooks

Several other Families of Coleoptera have species with lateral gill-like projections, e.g. Haliplidae and Dytiscidae. If larvae with lateral abdominal "gills" do not exactly fit couplets 15a, 16a, or 17a, they are probably Coleoptera.

15b End of abdomen without 4 recurved claws, but lateral filaments may be present. 16

16a With eight pair of 2-segmented lateral gills similar to Fig. 30, but abdomen ending with two false legs, each bearing two claws as in Fig. 29; size large, mature specimens from 3-7 cm in length (Fishflies and Dobsonflies). ....
Order Megaloptera, Family Corydalidae

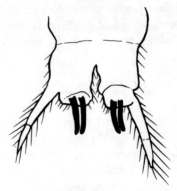

Figure 29

Figure 29 Corydalidae, detail of end of abdomen

Larvae of this family include the largest aquatic insects, some reaching three inches in length. The strong mandibles, jointed lateral gills, and a pair of anal prolegs, each bearing two claws, characterize this family. The biology of the common eastern species, *Corydalus cornuta*, has been studied extensively. Adults emerge in summer, mate, the female lays flat clusters of up to 1000 eggs on objects over the water. Larvae hatch and drop or crawl to the water, where they require 3 or more years to mature. This species prefers rapid streams. Other Corydalids live in swamps, ponds, and intermittent streams. All are formidable predators with powerful mandibles (See also Couplet 30b).

16b Size usually smaller and end of abdomen not as described; jointed lateral gills may be present. 17

17a Sides of abdomen with 7 pairs of 4-5 jointed lateral gills; end of abdomen with a single slender terminal filament (Fig. 30) (Alderflies). ....
...... **Order Megaloptera, Family Sialidae**

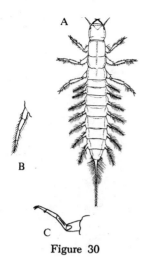

Figure 30

Figure 30 Sialidae. A. whole larva; B. detail of gill; C. detail of leg

The jointed lateral gills, strong mandibles, and single terminal filament at the end of the abdomen separate *Sialis,* the only genus in the family, from all other aquatic insects. When mature, they are about 2 cm in length. Larvae inhabit all types of aquatic habitats, from rivers and ponds to springs and seepage areas. They often burrow into soft substrates, or they may be found among gravel and rocks in streams. They are predatory and feed on anything they can overpower, including their own kind. Larval development requires 1 or 2 years, depending on climate. Larvae leave the water and dig a chamber in the soil for pupation. Adults emerge after about two weeks. (See also Couplet 30b).

17b Without the combination of a single terminal filament and jointed lateral gills; lateral unjointed filaments sometimes present (see text, Couplet 15a). 18

18a End of abdomen with a pair of lateral hooked claws on short or long prolegs (arrows, Figs. 31C, D, E); larvae often

in a case (Caddisflies, p. 101). .... Order Trichoptera

Figure 31 Trichoptera. A, B, F, G, H. various cases; C, D, E. anal claws of case building larvae (arrows)

Caddisflies are rather uniform and caterpillar-like in shape. A head capsule and 6 true thoracic legs are always present. There are never false legs on the middle abdominal segments (Fig. 26). Finger-like membranous gills are often present on the abdomen. Many species construct cases of various materials. Body length ranges from a few mm to 3 cm.

18b Larva never in a case; end of abdomen without hooked claws on prolegs; gills not as described above (Fig. 32) (Beetles, p. 121). .... Order Coleoptera

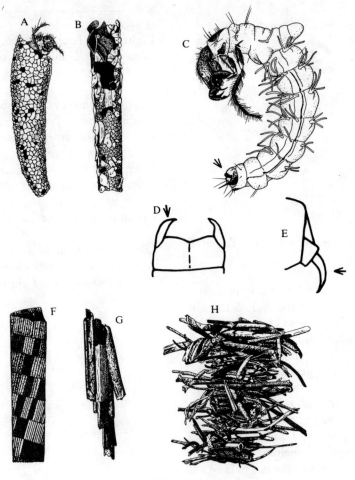

Figure 31

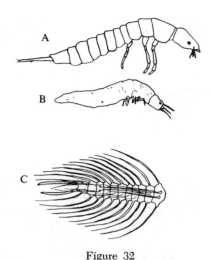

Figure 32

Figure 32 Coleoptera. A. Dytiscidae; B. Hydrophilidae; C. Haliplidae

Larvae of aquatic beetles are extremely varied in body structure and include oddities such as the Haliplidae illustrated in Fig. 32. Several unusual members are placed elsewhere in this key (Couplets 12a and 15a). The remainder are identified largely by elimination. They have a distinct head capsule and usually 6 true thoracic legs, but do not agree with the details in Couplets 13a-18a.

ADULT INSECTS

19a (from 1a) Front wings membranous and with distinct veins (Figs. 33, 35). 22

In some insects the wings and veins may be covered with hairs or scales. In such cases choose 19a.

19b Front wings not membranous but either hard and shell-like or leathery; veins indistinct or absent in the basal 1/2 (Figs. 15C, D, 17). 20

20a Mouthparts in the form of a segmented beak (Fig. 17); tips of wings usually membranous (Fig. 15D) (True Bugs, p. 93). .... Order Hemiptera

The segmented beak and leathery front wings separate adult Hemiptera from all other orders. In some families wingless adults are common. Body size ranges from a few mm to about 8 cm. Shape varies from oval to thread-like thinness.

20b Mouthparts chewing, front wings shell-like or leathery and not membranous at tips (Figs. 15C, 20). 21

21a Fore wing hard, shell-like and without veins; hind legs not enlarged for jumping (Fig. 15C) (Beetles, p. 128). .... Order Coleoptera

Adult beetles are easily recognized by the hardened and often shell-like front wings which cover the membranous hind wings, by the chewing mouthparts, and by the lack of cerci at the end of the abdomen. Body shape is rather variable and size ranges from a fraction of a mm to over an inch.

21b Front wings leathery; veins indistinct but present; hind legs modified for jumping (Fig. 20). .... Order Orthoptera, Grasshoppers and Crickets. See Couplet 7a.

22a Abdomen constricted where it joins the thorax; wing venation reduced (Fig. 33). .... Order Hymenoptera, Wasps

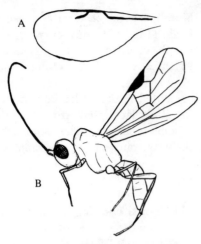

Figure 33

Figure 33 Hymenoptera. A. Wing with reduced veins; B. body form

Adults in the Order Hymenoptera are recognized by the two pairs of membranous, sparce-
ly veined wings, the hind pair being smaller
than the anterior pair, by the hard and well
sclerotized body, and in aquatic species, the
narrow-waisted abdomen. The aquatic species
all have the characteristic wasp-like appearance, but a few have reduced wings. No bees
or ants are aquatic.

Some spider wasps of the Family Pom-
pilidae are considered aquatic because they
specialize on aquatic spiders as prey. While
they actively crawl below the water surface
they have no obvious special aquatic adaptations.

Several different Ichneumonidae have
been observed under water and adults have
been reared from hosts such as aquatic Lepidoptera; the wasp families Braconidae and
Diaparidae are known to parasitize aquatic larvae of ephydrid flies. Other wasps are egg
parasites and are very tiny, some less than 1
mm in length. Egg parasite families and their
known hosts include Scelionidae on the eggs of
Gerridae (Water Striders) and Mymaridae
and Trichogrammatidae on the eggs of various
Coleoptera and Hemiptera. *Trichogramma*

semblidis is a well known Holarctic wasp that
can easily be reared from the eggs of *Sialis*
(Alder Flies).

**22b   Abdomen not constricted; wings not as
        in Fig. 33A or B.** **23**

**23a   Wings with scales which detach as a fine
        dust at the touch; mouthparts in the form
        of a coiled proboscis (Fig. 34) (Moths).**
        .... **Order Lepidoptera**

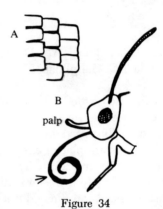

Figure 34

Figure 34 Lepidoptera. A. detail of wing
scales; B. head and coiled proboscis (arrow)

Over a dozen genera in the Family Pyralidae
are aquatic in the larval stage (see also Couplet 13a). In adults the labial palps are often
prominent, giving this family the common
name "snout moths". As adults, no special features separate them from terrestrial species
(except perhaps in the Schoenobiinae species
Acentria nivea where wingless females emerge
from the water only briefly to mate; winged
females for dispersal are also present in the
species). The Family Pyralidae is very large,
with over 1200 North American species including economically important species such
as corn borers, various sod web-worms, meal
moths, the wax moth, and many others. The
major subfamily, Nymphulinae contains about

44 (of a total of 66) species which are assumed to be aquatic in the larval stage. In general, these moths are 6-12 mm in body length, with a 2-4 cm wing-spread, and they are triangular in outline when at rest. In many there are black or metallic-markings on the posterior of the hind wings. Most have wavy markings across the wing. The excellent works of Lange (1956) and Monroe (1972-1973), the latter with color photos of North American species, should stimulate study of this interesting group. Adults are often collected at lights some distance from water, but they are also found in vegetation or among rocks near the larval habitat. Adults should be pinned and spread and preserved dry.

23b Wings without dust-like scales or coiled tube-like mouthparts. 24

24a Two pairs of wings present. 26

24b One pair of wings present. 25

25a Hind wings in form of halteres (club-like structures, h-Fig. 35B); cerci absent (True Flies, p. 149). .... **Order Diptera**

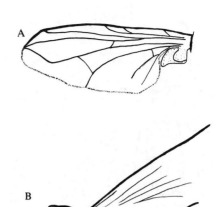

Figure 35

Figure 35 Diptera. A. Front wing; B. thorax with reduced hind wing or haltere (h)

True flies are unique among insects in having the rear wings modified into club-like halteres, and thus there is but one pair of functional wings. Crossveins are relatively few (Fig. 35A); body shape ranges from that of the mosquito to the housefly. Size ranges from minute, about 1 mm to over 25 mm in body length.

25b Halteres not present; long cerci on abdomen (Fig. 38) (Mayflies, p. 65). .... **Order Ephemeroptera, in part**

A few genera and species have lost the hind wings. See Couplet 28a.

26a Tarsi 3 segmented. 27

26b Tarsi 4 or 5 segmented. 28

27a Long, many segmented antennae; hind wing usually with an enlarged fan-like and area (a); cerci (c) present (but may be one-segmented) (Fig. 36) (Stoneflies, p. 89). ...... **Order Plecoptera**

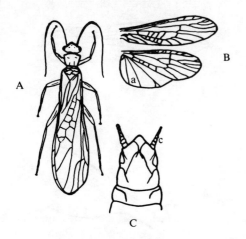

Figure 36

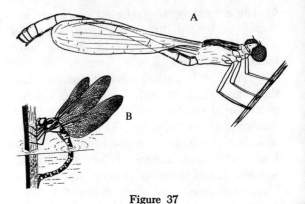

Figure 37

Figure 37 Odonata. A. Zygoptera (Damselfly); B. Anisoptera (Dragonfly)

Figure 36 A. Plecoptera; B. Wings with anal area (a); C. end of abdomen with cerci (c)

Stoneflies have long antennae, a moderate number of crossveins in the wings, distinct cerci on the abdomen (sometimes very short) and usually a folded fan-like area of the hind wing. Body size ranges from 6 mm to nearly 6 cm in length. Color ranges from black to green to nearly white. These are usually collected crawling on bridges or vegetation near flowing water. They are occasionally taken at lights.

27b Antennae short and bristle-like; hind wing without a fan-like area; cerci absent (Fig. 37) (Damsel and Dragonflies, p. 79). .... **Order Odonata**

The Order is easily recognized by the characteristic body shape, the short antennae, the lack of segmented abdominal cerci, the numerous crossveins in the wings, and the bristly front legs. Body length ranges from about 12 mm to over 6 cm. Many are brightly colored in life but fade when dried. These are strong, wide-ranging fliers and may be collected some distance from water. They are active in the day and are never taken at lights.

28a **Hind wing much shorter than the front, wings triangular in shape; long cerci present on the abdomen; antennae short (see also Couplet 25b) (Fig. 38) (Mayflies, p. 65). .... Order Ephemeroptera**

Figure 38

Figure 38   Ephemeroptera

This order is easily recognized by the long cerci on the abdomen and the somewhat triangular wings that have numerous crossveins. Mouthparts are greatly reduced since adults do not feed. Males, with almost no exceptions, have forceps, or clasping appendages at the end of the abdomen, greatly enlarged compound eyes, and much elongated front legs. They range in size from a few mm to 3 cm in body length. They are often collected at lights or in mating swarms at dusk near water.

28b   Hind wings not distinctly shorter than the front wings; antennae long. **29**

29a   Costal area of the front wing with few crossveins (Fig. 15A, 39); body shape as in Fig. 39 (Caddisflies, p. 112). .... **Order Trichoptera**

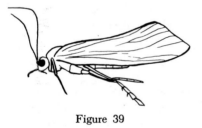

Figure 39

Figure 39   Trichoptera

These are moth-like insects with hairs on the wings (not detachable scales as in moths). They also lack the coiled-mouthparts of moths, but have reduced mouthparts, with only the palps being conspicuous. Body length ranges from a few mm to about 3 cm.

29b   Costal area of front wings with numerous crossveins (Fig. 40); body shape as in Figs. 41, 42). .... **Neuroptera, sensu latu.** **30**

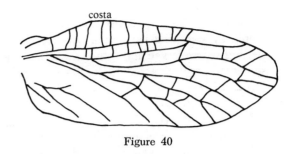

costa

Figure 40

Figure 40   Front wing of *Sialis* (Sialidae, Megaloptera)

Some authors include the three families which follow in the single Order Neuroptera. They are here divided into Megaloptera and Neuroptera, sensu strictu.

30a   Small grey or brown insects, about 1 cm in length (Fig. 41); hind wings similar in shape to the front wings (Spongilla Flies). .... **Sisyridae, Order Neuroptera, sensu strictu**

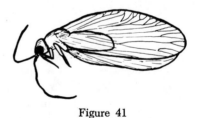

Figure 41

Figure 41   Neuroptera, Sisyridae

This family is the only aquatic member of the Order Neuroptera (sensu strictu). Adults have the appearance of a small brown lacewing. There are 2 genera (*Sisyra* and *Climacia*), each with 3 species, in North America. The Family is widely distributed, especially in the eastern half of N.A. and in the west from California to Alaska. *Climacia* has a number of crossveins in the outer half of the wing, while none are present in *Sisyra*. Adults have been observed feeding on nectar, pollen, and the eggs of other insects. Adults can often be collected by sweeping vegetation near ponds or streams where larvae and sponges occur. They sometimes are attracted to lights. (See also Couplet 14a).

30b   Larger insects; 15 mm to 6 cm in length, hind wings broader at the base than the front wings, anal area folded fan-wise when at rest (Alderflies and Dobsonflies). .... Order Megaloptera

Figure 42

Figure 42   Sialidae

In Sialidae, or alderflies, ocelli are absent, the body is less than 1 inch in length, and the 4th tarsal segment is bilobed (Fig. 42). Adult sialids are stocky, soft bodied, blackish insects. They have long antennae, an elongate head from side view, and they lack cerci on the abdomen. There are about 20 N.A. species which are widely distributed. All are in the genus *Sialis*. Adults are day-active insects, and females lay about 500 eggs in masses on objects overhanging water. These hatch in about 2 weeks. Eggs are often attacked by the wasp *Trichogramma semblidis*. (See also Couplet 17a).

In Corydalidae (Fishflies, Dobsonflies), three ocelli are present, the body is up to 6 or

7 cm in length and the 4th tarsal segment cylindrical (Fig. 43). The tarsi are 5-segmented, and the antenna is long and may be filiform (Fig. 43A) or pectinate (Fig. 43B). Some have a wing span of up to 15 cm.

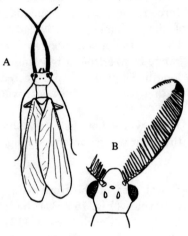

Figure 43

Figure 43   Corydalidae, A. Male with large tusks; B. detail of comb-like antenna

The family consists of about 20 species in half a dozen genera in the Neararctic region. Males of some species of *Corydalis* have hugh mandibles (Fig. 43). The group is widely distributed in the east and far west. Evans (1972) described scent glands in male *Corydalus cognatus* and *Orohermes* (= *Dysmiceohermes*) *crepusculus*. These may be functional in mating, or they may repel predators such as bats. Males apparently recognize females by sight and pursue them in flight. See also Couplet 16a.

Keys to Aquatic Insect Families

COLLEMBOLA
Springtails

These are tiny wingless arthropods, atypical among insects because of peculiarities of morphology and development, which are abundant and widespread and usually associated with damp habitats such as humus, leaf litter, moss or soil. Because of their habitat most species can be found associated with water under certain circumstances such as after rains; others are strictly aquatic. Several dozen species representing all families are considered by some authors to be aquatic.

Springtails derive the common name from their ability to jump by means of a leaping organ, the forked furculum (on the fourth abdominal segment) and the retinaculum (on the third segment) which holds it in place (Figs. 45, 50). The technical name Collembola is derived from an eversible sac-like organ, the collophore, also called the ventral tube, which is ventral on the first abdominal segment (Fig. 45).

Collembola are scavengers or vegetarians, feeding on algae, fungi, spores, lichens, living and dead plant material, or dead animals such as worms or snails. In their life cycle they do not undergo metamorphosis. Young Collembola hatch from the egg and have the appearance of small adults. Individual species are very widespread, and many North American species are either holarctic (North America, Europe, Asia) or cosmopolitan (worldwide).

Collembola can be stored in alcohol (95% preserves color pattern better than 80%) and for detailed study it is often necessary to mount specimens on slides.

1a Body elongate and distinctly segmented; thorax and abdomen not fused (Fig. 44). .... Suborder Arthropleona, 2

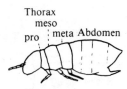

Figure 44

1b Body globular and without distinct segmentation dividing thorax from abdomen (Fig. 45). .... Suborder Symphypleona, Family Sminthuridae

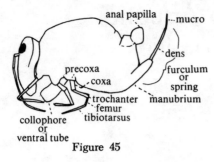

Figure 45

Figure 45   Sminthuridae

In this family the thorax and first 4 abdominal segments are fused into a single mass and abdominal segments 5 and 6 form an anal papilla which is separated from the rest of the body. The furculum or spring is always present. Eyes are present or absent. There are dozens of terrestrial North American species; about ten are considered aquatic. One of the best known is *Sminthurides aquaticus* which lives on the water surface or on vegetation such as duckweed. It is widespread in N.A. as well as in Europe and Japan. Males are 1/2 and females 1 mm in length. Color varies from rose-pink to deep purple.

2a **Prothorax reduced and without hairs or setae, often covered and hidden in dorsal view by the mesothorax (Figs. 46, 50). ...... Superfamily Entomobryoidea, 5**

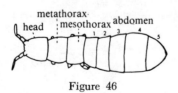

Figure 46

2b **Prothorax about equal in length to other thoracic segments and not hidden dorsally by mesothorax (Figs. 47-49). ....**
.... Superfamily Poduroidea, 3

3a **Small, elongate, eyeless, white insects which usually lack a furculum; pseudocelli present (chitinous rings on the integument at the base of the antenna or distributed over the body) (Fig. 47). ......**
.. Onychiuridae

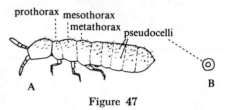

Figure 47

Figure 47   A. Body form of Onychiuridae; B. pseudocellus

The pseudocelli are often difficult to see in cleared specimens. This small family is usually associated with humus, soil and leaf litter but several species are considered to be aquatic.

3b **Usually dark-pigmented; usually with eyes and furculum; pseudocelli absent. ..**
... 4

4a **Spring long, so that when folded forward it would reach the hind pair of legs (Fig. 48). .... Poduridae**

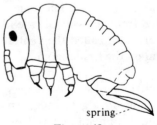

Figure 48

Figure 48   *Podura aquatica*

Represented by the single genus and species *Podura aquatica,* the family is widely distributed in N.A. and Eurasia. It reaches a length of 1.4 mm and is blue-black or red-brown in color. The antenna is shorter than the head and the furculum is long. Eyes are present. It dwells on the water surface or in trash along shore.

4b   Spring absent, or short, not reaching hind legs when folded forward (Fig. 49). .... **Hypogastruridae**

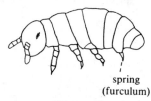

spring
(furculum)

Figure 49

Figure 49   Body form of Hypogastruridae

About a dozen species are associated with stagnant pools, marshes, shores of drainage ditches, and with the intertidal area of the seashore. In this family eyes may be present or absent, the mouthparts may be piercing and sucking or chewing, and the spring may be present or absent. Some have a distinctly "lumpy" appearance to the body.

5a   In dorsal view abdominal segment 4 much longer than segment 3, segment 4 often saddle-shaped; body often with large scales (Fig. 50). .... **Entomobryidae**

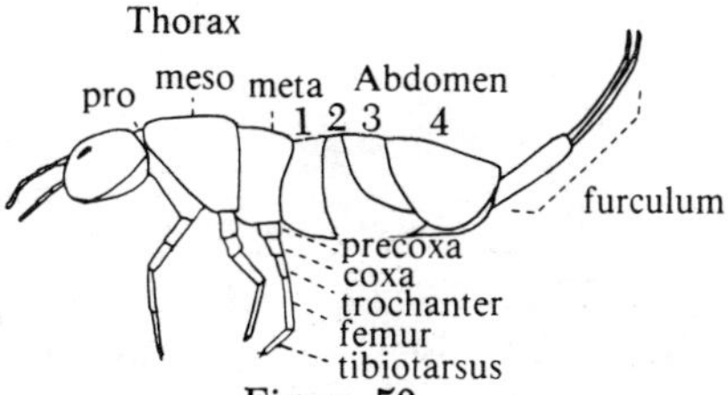

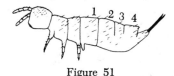

Thorax
pro meso meta Abdomen
1 2 3 4
furculum
precoxa
coxa
trochanter
femur
tibiotarsus
Figure 50

Figure 50   Body form of Entomobryidae

While this large group is usually associated with dry habitats such as the bark of dead trees or the underside of boards or stones, several species frequent the intertidal zone or stagnant pools.

5b   In dorsal view segment 4 never more than 1 1/2 times as long as segment 3; body with setae but not large scales (Fig. 51). .... **Isotomidae**

Figure 51

Figure 51   Body form of Isotomidae

This is a fairly large group of primarily terrestrial species but about half a dozen species frequent pools or the sea shore. Most well known of the aquatic species is *Isotomurus palustris.* This species is common in N.A., Mexico, Europe, Asia, and Australia. Body length is up to 3 mm; the typical color is yellow or greenish with a mid-dorsal stripe; eyes are present in this species.

EPHEMEROPTERA
Mayflies

(Note: Since mayfly nymphs are relatively easy to recognize to the genus level, several nymphal keys to genus are given. In other cases, subfamilies are given since no single set of couplets conveniently characterize an entire family.)

There are over 600 species and 60 genera of mayflies in North America. Adults are uniform in appearance (Figs. 38, 90B) but the group is very diverse in morphology in the nymphal stage (Fig. 19). Adults have triangular wings held vertically over the body when at rest; the front legs of the male are usually elongate. There are 2 or 3 long caudal filaments at the end of the abdomen in both sexes. The lateral pair are cerci; the median one, when present, is called the terminal filament. The same terminology is applied to nymphs.

Males have a pair of clasper-like forceps at the end of the abdomen (Fig. 90C). Above the forceps are the paired penes. In mating, which takes place during flight, the male flies under the female and reaches up with the long front legs and hooks them over the female wing bases. The end of the male abdomen bows upward and the male forceps (Fig. 90C) grasp the female abdomen. The male penes can then come in contact with the female genital opening. Copulation usually lasts but a few seconds.

Eggs of mayflies are diverse in shape and surface sculpture. Most females deposit eggs by flying just above the water surface and touching extruded egg masses onto the water. *Baetis* is an exception. Females crawl under the water, even in rapid streams, and deposit eggs in characteristic square patterns on submerged rocks (Fig. 2E).

Males often congregate in dense swarms during mating flights. These usually occur over water in the evening, but in some cases swarms are found up to 50 feet above ground in bright sunshine. Mayflies of some species are attracted to lights. Few are in mountainous or northern areas where nights are cool. During peak emergence periods the burrowing mayflies (especially *Hexagenia*) may emerge from rivers and lakes in huge numbers, even to the point of interfering with traffic because the drifts and piles of mayfly bodies may be several feet deep.

The subimago stage is unique to mayflies. This is a winged stage which emerges from the nymph, but which must moult once more, usually after a period of about 1 day. This moult produces the sexually mature adult. Subimagos have a dull gray color, the genitalia and legs are incompletely developed, and the wings have a fringe of hairs. Subimagos can usually be identified to family but seldom to genus or species. Subimagos should not be touched on the wings when collected (it will prevent successful moulting), but should be carefully coaxed into a small box of some type which contains a leaf for humidity, and they will transform into the taxonomically useful adult stage.

Nymphs of mayflies occur in all types of waters. The greatest diversity and abundance is found in streams and rivers. Almost every genus is distinctive in appearance. Nymphs can be broadly categorized. Only a small number are predatory, feeding primarily on Chironomidae (*Anepeorus, Pseudiron, Analetris,* and *Spinadis*). Most feed on plant material or detritus. Many genera burrow in river or lake bottoms (Ephemeroidea). *Lachlania, Isonychia,* and *Ametropus* filter particles from the water current; a number are streamlined swimmers which graze on the substrate surface (*Baetis, Ameletus*); Heptageniinae and many Leptophlebiidae are flattened dorsoventrally and live on or under rocks. Many are sprawlers which move over the substrate (*Baetisca, Tricorythodes,* some *Ephemerella*). There are many intermediates, making these nymphs among the most diverse in structure among the aquatic insects.

All stages are normally preserved in 70% alcohol. Pinned adults are extremely fragile. For study of adults it is usually best to carefully remove one pair of wings with fine pins and forceps and mount them on a slide (Fig. 14B) so that the veins can be easily seen. A hand lens should be sufficient to allow family identification.

EPHEMEROPTERA NYMPHS*

1a   Front corners of head and prothorax crowned with dense patches of hair-like setae (Fig. 52). .... Behningiidae

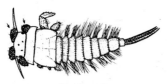

Figure 52

Figure 52   Behningiidae (*Dolania americana*)

Dolania americana, the only North American representative of this family, is found from South Carolina to Florida, where it burrows in the sand of warm rivers and streams. The species is rare but may be abundant in favorable locations. The morphology is very unusual, with ventral fringed gills, dorsal crowns of golden setae on the head and prothorax, setae generally distributed over the body, and legs highly modified for burrowing in loose sand. A ventral view of the mouthparts and front legs is strongly reminiscent of what is seen in the crablike Crustacea, with numerous finger-like palps used to sort and manipulate detritus which apparently is the food of these nymphs. Examination of the gut contents of a nymph

in my collection yielded no animal remains, but only unidentifiable debris. However, these nymphs are said to be predators.

1b   Front corner of head and prothorax without dense patches of setae. 2

2a   Dorsal thorax expanded to cover the first 5 or 6 abdominal segments and the gills; often with dorsal or lateral spines (Figs. 53, S). .... Baetiscidae

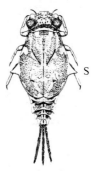

Figure 53

Figure 53   Baetiscidae (*Baetisca*)

This family consists of the single widespread genus *Baetisca* which has about a dozen described species restricted to the New World. The morphology is unique, nymphs being hump-backed and spiny as a result of the expanded thoracic notum. The family has no close relatives, and the only similarly constructed mayflies are the Old World Prosopistomatidae, which are smooth and oval rather than spiny. Nymphs prefer flowing water or wave-washed lake shores and are found in situations ranging from the undersides of submerged boards, pebbly riffles, silty areas to plant stems. The degree of spinyness varies

*This key does not include *Spinadis* Edmunds and Jensen (Heptageniidae) known from large rivers in Georgia, Indiana, and Wisconsin, which resembles Fig. 75 in body shape, but has tubercles dorsally on the head, thorax and the abdominal segments.

considerably among species; all appear to be grazers or detritivores. See adult key for distribution.

2b **Thorax not expanded; appearance not as in Fig. 53.** 3

3a **Mandibles (m) enlarged to form stout tusks visible from dorsal view (Fig. 54).** 4

Figure 54

Figure 54 Dorsal view of head with tusks (mandibles), m

3b **Mandibles not visible from dorsal view.** 8

4a **Gills curved above abdomen, body not flattened dorsoventrally; forelegs fossorial, modified for digging and bent close to body (Fig. 55).** 6

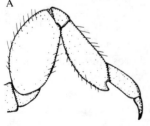

Figure 55

Figure 55   A. Leg modified for digging (fossorial); B. Dorsal view of abdomen with gills (g), diagrammatic

4b **Gills extended laterally; body somewhat flattened dorsoventrally; forelegs grasping (Fig. 57C).** 5

5a **Gills without fringes on edges (Fig. 56).** .... **Paraleptophlebia** (in part) **Leptophlebiidae**

Figure 56

Figure 56 *Paraleptophlebia*. A. head and mandibles; b. gill

Several western species of *Paraleptophlebia* develop large smooth sickle-shaped mandibular tusks in the late instars. These species range from Alberta, Montana and Utah to the west coast. Included are *P. bicornuta, P. packi, P. helena,* and *P. zayante.* They are usually found on or under stones in slow-moving water.

5b **Gills fringed on edges (Fig. 57).** .... **Potamanthidae**

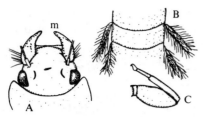

Figure 57

Figure 57   Potamanthidae (*Potamanthus*), A. head and mandibles (m); B. abdomen and lateral gills; C. front legs

Potamanthus, the North American representative of this family, consists of 8 species which inhabit the eastern half of N.A. These are the sprawling Ephemeroidea (most members of this superfamily are burrowers) and are found on rocks and gravel or silty bottoms in flowing water. They may burrow in the early instars. Larval morphology corresponds to the non-burrowing habit: gills are lateral rather than recurved over the abdomen; the legs are outspread and not modified for burrowing, and the body is somewhat flattened. While relatively common and widespread, the biology of the group is poorly known.

6a   Head with a rounded or forked projection between antennae (Fig. 58). 7

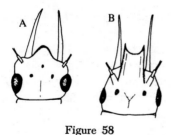

Figure 58

Figure 58   A. *Hexagenia;* B. *Ephemera*

6b   Without an anterior projection on the front of the head between the mandibles (Fig. 59). ....
.... Campsurinae, Polymitarcyidae

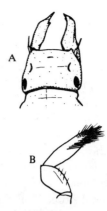

Figure 59

Figure 59   Polymitarcyidae (*Tortopus*) A. head and mandibles; B. front leg

This subfamily consists of two N.A. genera, *Tortopus* and *Campsurus*. Nymphs are characterized by the fringed gills curved dorsally over the abdomen, the lack of a projection on the front of the head, by stout convergent mandibular tusks which have a distinct tooth on the inner edge, and by the fore tarsi partly fused to the tibiae. Little is known of the biology except that *Tortopus* burrows in clay banks in rivers, and frequently makes dense honeycomb burrows in this stable substrate. *Campsurus* lives in mud and silt in streams or rivers.

7a   From side view mandibular tusks curve downward; tusks with a rough dorsal surface (Fig. 60). ....
.... Polymitarcyinae, Polymitarcyidae

Figure 60

Figure 60 Polymitarcyidae (*Ephoron*), side view of head

The subfamily consists of the single genus *Ephoron*, with two N.A. species. *E. album* and *E. leucon* burrow in gravily or clay substrates of streams, rivers and lakes and feed on detritus, which is filtered from the water or gathered from the substrate. The genus is recognized by the fringed gills which curve dorsally over the abdomen, the presence of a small rounded frontal process on the head between the antennae, and mandibular tusks which are convergent and downcurved apically and which have conspicuous small spines on the dorsal surface. Mature larvae are about 2 cm long and generally white with dark markings. Body shape resembles Fig. 61A.

7b **From side view, mandibular tusks curve upward (Fig. 61B).** .... **Ephemeridae**

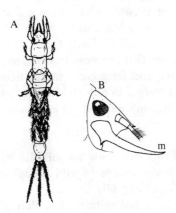

Figure 61

Figure 61 Ephemeridae; A. whole nymph; B. side view of head with mandible (m)

This family contains four North American genera (*Hexagenia, Ephemera, Litobrancha* and *Pentagenia*). All are large mayflies which burrow in silt and sand or clay in streams, rivers, and lakes. They are often important as food for game fish. Up to three years are re-

quired to complete nymphal development. The family is characterized by fringed gills which recurve dorsally over the abdomen, the presence of a frontal process on the head, and by upturned mandibular tusks. *Litobrancha* and *Hexagenia* have a dome-shaped frontal process (Fig. 58A); the former lacks whorls of setae on the antennae but these are present in *Hexagenia*. The frontal process is bifurcate in *Ephemera* and *Pentagenia* (Fig. 59B); the dorsolateral angle of the mandibular tusks is smooth in *Ephemera* but toothed and rough in the latter. *Pentagenia* is sometimes placed in the family Palingeniidae.

8a **Femur and tibia of the front leg with conspicuous, inwardly-directed fringes of long hairs (Fig. 62).** **9**

Figure 62

Figure 62 Fore leg of *Lachlania*

8b **Front femur and tibia without long, inwardly-directed hairs; small bristles or fine hairs may be present but not arranged as in Fig. 62.** **10**

9a **Six pairs of gills visible from dorsal view, gills of segment 1 ventral, other six pairs of gills lateral or dorsal (Fig. 63).** .... **Oligoneuriidae**

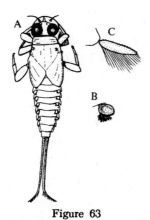

Figure 63

Figure 63 Oligoneuriidae. A. Nymph of *Lachlania*; B. gill of *Lachlania*; C. gill of *Homoeoneuria*

Nymphs of this primarily tropical family inhabit large warm rivers and streams, and while locally abundant, they are rarely collected. *Lachlania* (with two caudal filaments) and *Homoeoneuria* (with three) are the genera found in North America. Both have a tuft of gills at the bases of the maxillae. *Lachlania* grasps roots and other objects in flowing water and filters food from passing water with the long setae on the front legs, similar to *Isonychia* of the family Siphlonuridae. *Homoeoneuria* burrows in loose sand and feeds on material caught in the hairs on the legs. *Lachlania* has both a platelike and a fibrilliform portion to the gills (Fig. 63B) while *Homoeoneuria* lacks the latter (Fig. 63C).

9b **Seven pairs of gills visible from dorsal view; all seven pairs attached in a similar way, gills of segment 1 not ventral (Fig. 64).** ....
.. *Isonychia*, **Isonychinae, Siphlonuridae**

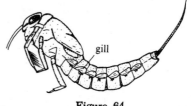

Figure 64

Figure 64 Siphlonuridae (*Isonychia*)

The subfamily contains the single genus *Isonychia*. While nymphs of this genus have hair fringes used for filter feeding in flowing water as do *Lachlania* of Oligoneuriidae, their relationship is distant. Another peculiarity of *Isonychia* shared with the above is the presence of tufted filamentous gills on the maxillae and fore-coxae. The genus is common and **widespread** and in favorable situations they are very abundant.

10a **Gills on segment two enlarged, rectangular, extending over abdomen to meet at midline and covering other gills (operculate) (Fig. 65).** **11**

(*Note*: To determine the segment a gill is attached to, begin with the last or 10th segment from dorsal view and count forward. Gills are attached to the postero-lateral corners of the segment.)

Figure 65

10b Gills on segment two (or three) may be operculate but are not rectangular and meeting at the midline (Figs. 68 and 69). **12**

11a Operculate gills fused at midline; prothorax (p) with extended and sometimes sharp antero-lateral angles; mesothorax (m) with rounded lobes at antero-lateral angles; mature nymphs 8-15 mm body length; hind wing bud present (Fig. 66). .... **Neoephemeridae**

Figure 66

Figure 66 Neoephemerdae (*Neoephemera*); p. prothorax; m. mesothorax

There is a single North American genus, *Neoephemera,* with 4 species which range from Quebec to Michigan and south to Florida.

While adult venation resembles Ephemeridae, nymphal morphology is similar to Caenidae. The gills of abdominal segment 1 are small and filiform (thread-like) (see arrow, Fig. 67, and compare, Fig. 66); the second pair are operculate and cover the other gills. Nymphs are rather infrequently collected. They inhabit streams where they can be found in moss or tangles of roots and branches along the shore. Nymphal development takes about 1 year. Plant epidermis and algae cells have been found in the gut.

11b Operculate gills not fused at midline; prothorax not as well developed at front corners and without a distinct lobe at the antero-lateral angle of the mesothorax; mature nymphs 3-7 mm long; hind wing buds absent (Fig. 67). .... **Caenidae**

Figure 67

Figure 67 Caenidae (*Caenis*); arrow, gill 1

Caenis and *Brachycercus,* with about 1 dozen and 1/2 dozen species respectively, constitute the family in North America. *Brachycercus* has a conspicuous median tubercle on head and is therefore easily separated from *Caenis. Brachycercus* inhabit large rivers where they burrow through sand and sediment sorting out edible particles from the debris. *Caenis* are characteristically found in ponds and are often extremely abundant. They feed on organic material. In both, the respiratory gills are protected from debris by the operculate gills of segment two.

12a   Gills on segment 1 large, resembling those on segment 3. 14

12b   Gills on segment 1 absent (Fig. 68) or reduced to inconspicuous thread-like structures (as in Fig. 67, arrow). 13

13a   Gills lacking on abdominal segment 2 (Figs. 68, 87, 88). .... p. 62, Ephemerellidae

Figure 68

Figure 68   Ephemerellidae; gill of segment 3

This family contains the single genus *Ephemerella* and includes about 75 species which are widespread and common in the flowing waters of North America. Although the group is variable in body proportions and in the presence or absence of features such as spines, they are quite uniform in basic structure and habits. They are stream and river forms which cling to wood or rock substrates and browse on algae and similar material. The genus has been divided into a number of subgenera which are covered in this key. These groups are not always easily separated.

13b   Gills present on segment 2, operculate, triangular or oval and not meeting at the mid-line (Fig. 69). .... Tricorythidae

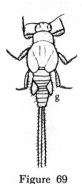

Figure 69

Figure 69   Tricorythidae (*Tricorythodes*)

This family is represented by two genera, the widespread and common genus *Tricorythodes* with about a dozen species, and the rarely collected and mainly South American and Mexican genus *Leptohyphes*. In *Tricorythodes* the operculate gills are triangular while they are oval-elongate in *Leptohyphes*. Nymphs of this family superficially resemble Caenidae. *Tricorythodes* typically occur in moderate to gently flowing water and feed on algae and debris.

14a   Claws of front legs split by a cleft; claws of middle and hind legs long and slender (Fig. 70). .... Metretopodidae

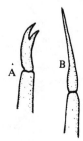

Figure 70

Figure 70 Claws of Metretopodidae. A. front; b. middle

This family contains the genera *Siphloplecton* and *Metretopus*. The former is widespread in eastern temperate regions of North America. *Metretopus* is a northern form and is found across Canada, in Michigan and Maine, and Europe. Both are found in flowing water, usually medium to large-sized rivers. Body shape resembles Fig. 64.

14b Claws of front legs not split but may have teeth (Fig. 71). **15**

Figure 71

Figure 71 Mayfly tarsal claws

15a Claws of all legs short, claws much shorter than the tibia and tarsus combined (Fig. 72). **18**

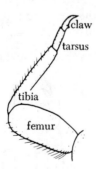

claw
tarsus
tibia
femur

Figure 72

15b Claws of middle and hind legs very long, often as long as the tibia and tarsus combined (Figs. 73, 74, 75, 76). **16**

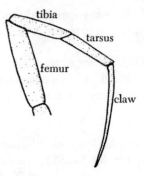

tibia
tarsus
femur
claw

Figure 73

16a Claws of front legs with hairs; coxa of front leg with a large lobe (arrow, Fig. 74C). .... ***Ametropus*, Ametropodidae**

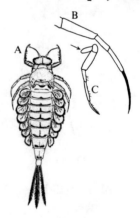

Figure 74

Figure 74 Ametropodidae (*Ametropus*). A. Nymph; B. middle leg; C. front leg; arrow—coxal lobe

Ametropus, the only genus in the family, inhabits sand bars and sedimented areas in rivers from Saskatchewan to Utah and west to British Columbia and Oregon. Distribution is very spotty but nymphs are locally abundant. Nymphs grip the substrate with the 4 posterior legs, and after settling into the sand, collect food material from the sand and current with the hairy front tarsi and claws.

16b Claws of front legs without hairs; fore-coxa without a lobe. 17

17a Gills elongate with a filament at mid-length and with a basal tuft of filaments (Fig. 75B); body with long, crab-like legs as in Fig. 75A. ....
.... Pseudironinae, Heptageniidae

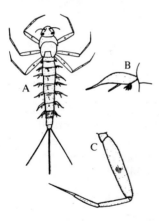

Figure 75

Figure 75 Heptageniidae (*Pseudiron*). A. Nymph; B. gill; C. leg

This subfamily was formerly placed in Ametropodidae because of the long tarsal claws. Recent studies however place it in Heptageniidae because of adult characters. There is but one genus (*Pseudiron*) with two species (*P.*

meridionalis and *P. centralis*). Nymphs are carnivorous, feeding on Chironomidae and other prey which can be captured on the current-swept, submerged sand bars of large rivers which they inhabit. They are well adapted and mobile on this shifting habitat. The long claws, tibia and tarsus plunge into the sand, working like ice tongs to grip the sand. They move crab-like over the sand. While considered to be rare, they are widespread and locally abundant.

17b Gills round and body shape resembling Fig. 76 (*Analetris*), or gills elongate and curved with a ventral flap and abdominal segments 1-9 with a median hooked tubercle both dorsally and ventrally (*Acanthametropus*). ....
.... Acanthametropodinae, Siphlonuridae

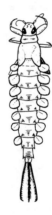

Figure 76

Figure 76 Siphlonuridae, (*Analetris*)

This is a rare group of mayflies which is carnivorous in the nymphal stage. The only well-known species is *Analetris eximia* which inhabits streams and rivers from Saskatchewan to Wyoming and Utah. It is abundant under favorable conditions. They are dorsoventrally flattened and are excellent swimmers. *Acanthametropus* (= *Metreturus*) has not been collected recently and is known from only a few nymphs taken in Georgia and Illinois.

18a Body flattened dorsoventrally, eyes and antennae on top of head, most mouthparts hidden from dorsal view (Fig. 77). **19**

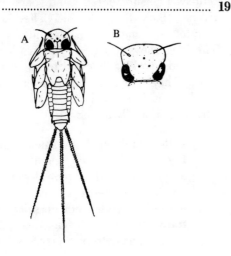

Figure 77

Figure 77   A. Heptageniidae; B. dorsal view of head

18b Body not flattened dorsoventrally; eyes and antennae on sides of head (Fig. 78). **21**

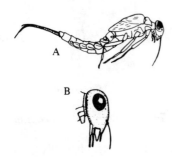

Figure 78

Figure 78   A. Body and B. head, lateral view

19a Gills attached ventrally, consisting of hook-like lamellae and a tuft of filaments which are longer than the lamellae (Fig. 79C); head from ventral view with tong-

like labia palps (Fig. 79B). .... **Anepeorinae, Heptageniidae**

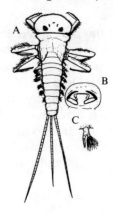

Figure 79

Figure 79   *Anepeorus*. A. Whole nymph; B. ventral view of head with labium; C. abdominal gill

This subfamily consists of the rare and unusual genus *Anepeorus*. It is distributed from Saskatchewan to Illinois and Iowa. These nymphs are carnivorous and are clearly distinct from other Heptageniidae. In the Saskatchewan River they inhabit deep swift water where they can be dislodged from rubble.

19b Gills attached dorsally or laterally on the abdomen and not agreeing exactly with the above in structure; labium not tong-like. **20**

20a Gills as in Fig. 80, consist of a single plate usually with a tuft of filaments at the base; body shape as in Fig. 77A. .... **Heptageniidae, Heptageniinae and Arthropleinae, 23**

Figure 80

Figure 80   Gills of Heptageniidae

20b   Gills as in Fig. 81, either forked, or consists of a cluster of filaments, or are bi-
lamellate with fringed margins; if they
consist of double plates which terminate
in a point, then gill 1 is forked or of a single filament; body less flattened than
Fig. 77A. Page 63. .... **Leptophlebiidae**

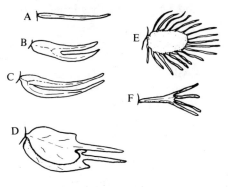

Figure 81

Figure 81   Gills of Leptophlebiidae

This is a widespread and common family consisting of about half a dozen genera and 60 species. Nymphs are somewhat dorsoventrally
flattened but can be separated from the Heptageniidae by the gills. The family inhabits
both still and flowing water and they feed on
plant material. Genera are separated in the
key on page 63.

21a   Gills single or double plates, sometimes
with a recurved flap and sometimes
pointed at the tip; none forked (Fig. 82).
.... 22

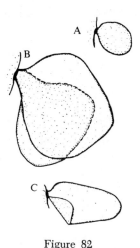

Figure 82

Figure 82   Mayfly gills

21b   Gills as in Fig. 81, either forked, or consists of a cluster of filaments, or are bi-
lamellate with a fringed margin; if they
consist of double plates which terminate
in a point, then gill 1 is forked or of a single filament (Fig. 81), page 63, (see al-
so Couplet 20b). ....
.... some **Leptophlebiidae**

Because of the variability in this family, it
appears twice in the key.

22a   Antennae more than twice as long as
head (Fig. 83); postero-lateral margin of
abdominal segments 8 and 9 not at all or
weakly produced as spines, page 64. ....
.... **Baetidae**

Figure 83

Figure 83   Antenna of Baetidae

Nymphs of this family are among the most common mayflies. *Callibaetis* is a very common pond mayfly found throughout North America. *Baetis* is among the most common stream mayflies. Other North American genera include *Apobaetis, Baetodes, Centroptilum, Cloeon, Dactylobaetis, Neocloeon, Paracloeodes,* and *Pseudocloeon.* Many members of the family are difficult to identify to the genus and species level, but a key to genera is given on page 64.

22b Antennae less than twice as long as head; postero-lateral margins of abdominal segments 8 and 9 produced as distinct spines (Fig. 84). ....
.... **Siphlonurinae, Siphlonuridae**

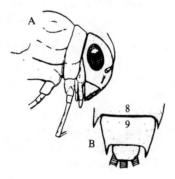

Figure 84

Figure 84   A. Antenna of Siphlonuridae; B. postero-lateral spines of abdominal segments 8 and 9

This subfamily consists of 5 genera, two of which (*Ameletus* and *Siphlonurus*) are very common and widespread. Most are minnow-like stream inhabitants which feed on plant material. *Siphlonurus* (often found in ponds) and *Edmundsius* have gills formed of double lamellae on segments 1 and 2 (Fig. 82B); *Edmundsius* has mid and hind tarsal claws about twice as long as the claws of the front legs; in *Siphlonurus* all tarsal claws are about equal in length.

The other genera have gills of a single lamella (Fig. 82A). *Ameletus* has 1 or 2 sclerotized bands on the gills and a comb of pectinate spines on the maxillae. *Siphlonisca* has abdominal segments 5-9 greatly expanded laterally, and *Parameletus* has labial palpi which form a pincer-like process.

23a Maxillary palpi enlarged into long sweeping organs about twice as long as the head (Fig. 85). ....
.... **Arthropleinae, Heptageniidae**

Figure 85

Figure 85 Maxillary palp of *Arthroplea*. A. Head region (much distended); B. maxilla and palp

This subfamily contains the single genus *Arthroplea* which is distinct because of the long maxillary palpi that are used in straining the water and sweeping food to the mouth. There is but one species in North America, *A. bipunctata*, which is known from Quebec, Ontario, Mass., Me., and N. H. A similar species is found in Europe. Nymphs inhabit bogs and ponds.

23b **Maxillary palpi not elongated as above (Fig. 86), page 61.** ....
.... **Heptageniinae, Heptageniidae**

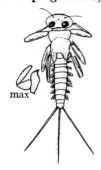

max

Figure 86

Figure 86 Heptageniinae and maxilla

The Heptageniinae are a fairly well defined group of dorso-ventrally flattened, primarily stream-inhabiting mayflies. Odd members of the family have been placed in their own subfamilies (see Anepeorinae, Pseudironinae, and Arthropleinae, and footnote (page 49) in this key). The remaining genera are separated in the key on page 61.

Members of this group are found almost everywhere there is flowing water. Some have been found on shores of mountain or northern lakes. All feed on algae, detritus or plant tissue. The group contains over 100 species.

KEY TO HEPTAGENIINAE GENERA

1a Nymph with 2 caudal filaments only. ....
.... *Epeorus*

Found in cool rapid streams in the west and northeast, this common genus contains about 20 species. The former subgenus, *Ironodes* (with paired submedian spines on the dorsum of the abdominal segments) is now given full generic status. It is western in distribution.

1b With 3 caudal filaments. 2

2a From ventral view, gills form a sucker-like disc. .... *Rhithrogena*

There are about 20 species in this genus and they live in cool, clear, flowing water primarily in the west and northeast.

2b Gills lateral or dorsal and do not form a sucker-like disc. 3

3a Gills of 7th abdominal segment in the form of a tapered filament, somewhat spinelike in appearance. .... *Stenonema*

This is a common eastern and midwestern genus which is uncommon in the west. There are nearly 30 species and nymphs occupy a wide range of flowing water habitats. *Stenacron* has recently been separated from this genus and can be recognized by the pointed rather than rounded or truncate gills on segments 1-6.

3b Gills of 7th segment similar to those on the preceeding segments, not spinelike. 4

4a Fibrilliform portion of gills reduced to a few threads or completely lacking; anterior margin of head with a median emargination or indentation. .... *Cinygmula*

This is a common western genus containing about 10 stream-inhabiting species.

4b Fibrilliform portion of gills larger; front of head without a distinct emargination. 5

5a   Labrum 1/3 the width of the anterior margin of the head and gills on segment one distinctly smaller than gills on segment two. .... *Cinygma*

There are but 3 species in this western genus; all live in streams and are often found on decaying wood.

5b   Labrum 2/3 to 3/4 the width of the anterior margin of the head; gills on segment one not distinctly smaller than gills on segment 2. .... *Heptagenia*

This is a common and widespread genus. There are over 30 species and they live in a wide range of flowing water habitats.

KEY TO EPHEMERELLIDAE SUBGENERA

This Family consists of the single genus *Ephemerella* which is divided into the subgenera indicated in the following key.

1a   Gills present on segments 3-7. 2

Five pair are present, but the 5th may be small and covered by the 4th (Fig. 87, arrow).

1b   Gills present on segments 4-7 only. 5

Gill on 4 may be enlarged (Fig. 88) and must be lifted to count the gills on the posterior segments.

2a   Lateral cerci 1/4-3/4 as long as the median terminal filament (Fig. 87). .... *Caudatella*

Figure 87

Figure 87   *Ephemerella* (*Caudatella*);   5th gill

Six species are included in this subgenus. All are from western North America, ranging from Alberta and Montana to British Columbia and south to California.

2b   Lateral cerci and median terminal filaments about equal in length (Fig. 88). 3

3a   Tubercles present on the leading edge of the fore-femora *or* with a ventral sucker disc on abdominal segments 3-8 *or* head, thorax and abdomen with well-developed spine-like tubercles. .... *Drunella*

Fifteen species are included in this structurally diverse group. All but three species have large tubercles on the front edge of the fore-femur. While members of *Ephemerella* also may have small dorsal tubercles, these may be extremely large in *Drunella*. *E. doddsi* and *E. pelosa* have complete or partial ventral sucker discs on the abdomen. The group is widespread and common in North America wherever there is flowing water.

3b   Tubercules wanting on the leading edge of the front femora, without a ventral adhesive disc, and with only small tubercles or none on head, thorax and abdomen. 4

4a Caudal filaments with hair-like interseg-mental setae *or* abdomen with small tuber-cles dorsally on segments 2-7; legs long and thin. .... *Ephemerella*

This subgenus contains 25 species and is common and widely distributed in North America.

4b Caudal filament with only sparce or no in-tersegmental setae; abdomen without dorsal tubercles. .... *Serratella*

Serratella includes 13 North American spe-cies and is widespread and common.

5a Tarsal claws without denticles. 6

5b Tarsal claws with denticles. 7

6a Apex of each femur terminating in a sharp spine, western N.A. (Fig. 88). .... *Timpanoga*

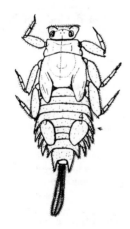

Figure 88

Figure 88 *Ephemerella* (*Timpanoga*); 4th abdominal segment

This taxon includes only the very distinctive spe-cies, *Ephemerella hecuba*, which is widespread in the mountainous regions of western North Amer-ica.

6b Apex of each femur without a spine, eastern North America. .... *Dannella*

Two species, *E. simplex* and *E. lita* form this group and they are widely distributed in the eastern half of North America.

7a Segment 9 of abdomen longer than 8, gills on segment 4 semi-operculate. .... *Eurylophella*

Twelve species make up this group in North America. All but *E. lodi* Mayo are restricted to the eastern half of the continent. *E. lodi* ranges from British Columbia to California.

7b Segment 9 of abdomen about equal in length to 8, gills of segment 4 not operculate. .... *Attenuatella*

Attenuatella comprises four species and is widespread in North America.

KEY TO LEPTOPHLEBIIDAE GENERA

1a Gills bordered with filaments (Fig. 81E) and labrum as broad as head. .... *Traverella*

Four species of this warm river mayfly are found in the interior west from Saskatche-wan and Alberta to New Mexico and Arizona.

1b Labrum narrower and gills not as in Fig. 81E. 2

2a Gills 2-6 consists of clusters of slender fila-ments (Fig. 81F). .... *Habrophlebia*

These are very small, about 5 mm when mature, and the single species, *H. vibrans*, is found from eastern Canada to Florida, where they live in streams.

2b Gills not as above but as Fig. 81 A, B, C, or D. 3

3a Gills on segment 1 different from those on segments 2-7. 4

3b Gills on segment 1 similar to those on seg-ments 2-7. .... *Para-leptophlebia, Habrophlebiodes, Thraulodes*

These genera are not always easily separated. *Paraleptophlebia* is the largest and probably most widespread genus in the family, with over 30 species. They are usually found in flowing water. The body is only slightly flattened and the gills are as in Figs. 81B or C, i.e. divided either to the base or only half way to the base. Abdominal terga 1-10 have spinules on the posterior margins. Body length is up to 10 mm. *Habrophlebiodes* is eastern in distribution (Quebec and Ontario to Illinois, Oklahoma and Florida). Nymphs are small (4-6 mm), the anterior margin of the labrum is deeply cleft (shallow in *Paraleptophlebia*), gills are as in Fig. 81B, spinules are on abdominal terga 6-10 only. They live in streams and there are 4 species. Five species of *Thraulodes* are found in the southwest (Arizona, N. Mex., Texas) and the genus is very common in Central America. All gills are double (Fig. 81D) but the apex is evenly tapered. In contrast to the above genera, the head is squarish and flattened, the body is more flattened, and the mouthparts are directed anteriorly rather than ventrally. *Thraulodes* lives in streams.

4a Gills on segment 1 forked (Fig. 81B); gills on 2-7 similar to Fig. 81D. .... *Leptophlebia*

There are about 10 species widely distributed in North America. They are found in both ponds and flowing water.

4b Gills on segment 1 a single filament (Fig. 81A) and gills on 2-7 similar to Fig. 81D. .... *Choroterpes*

This genus contains about 10 species which are widespread in North America.

KEY TO BAETIDAE GENERA

Many taxonomic difficulties remain in this family, but most nymphs can be identified with this key. The state of knowledge in Baetidae is far from satisfactory even though they are among the most common mayflies, especially *Baetis* in streams and *Callibaetis* in ponds.

1a Tarsal claws modified, flattened and spatulate with a saw-like outer edge (Fig. 89). .... *Dactylobateis*

Figure 89

Figure 89   Claw of *Dactylobaetis*

This genus is mainly Central and South American but 2 species have been reported from the west, ranging from Utah and Saskatchewan to Oregon and California. Uncommon.

1b Claws pointed (Figs. 71, 72), may have a toothed edge but not modified as Fig. 89. .... **2**

2a Gills consist of double lamellae *or* of single lamellae with a recurved flap (Figs. 82B, C). .... *Callibaetis, Cloeon, Centroptilum* **(in part)**

These are sometimes difficult to separate. *Cloeon* lacks metathoracic wing pads but these are present in the other two. *Callibaetis* has double gills on segments 1 and 2 and the remaining gills have a ventral recurved flap. Many *Centroptilum* have simple gills, but some have a basal dorsal flap on the gills. All are widespread and common but poorly studied.

2b Gills a simple, flat lamella (Fig. 31A). **3**

3a With but 2 well-developed caudal filaments. .... *Heteroclocon, Baetodes, Baetis* **(in part)**, *Pseudocloeon*

Baetodes (mainly South America but ranging to Texas and Arizona, found in streams) has gills on segments 1-5 only while the others have gills on segments 1-7. *Pseudocloeon* lacks metathoracic wing pads. They live in streams and the body is somewhat flattened. Metathoracic wing pads are present (but often minute) in *Baetis* and *Heteroclocon*. Most *Baetis* have 3 tails. In *Hetero-*

cloeon the division between the 2nd and 3rd segments of the labial palp is indistinct, but is distinct in *Baetis*. The latter two also live in streams.

3b With 3 well-developed caudal filaments. .. 4

4a Middle tail shorter and thinner than the lateral tails (Fig. 78A). .... *Baetis* (in part)

See also 3a. This is the most common and widespread genus with over 50 species. They live primarily in streams.

4b Middle tail about equal in size to the lateral tails. **5**

5a Gills with poorly developed lateral branches on the trachea. ...... *Paracloeodes, Apobaetis*

These are uncommon and have been reported only from California. *Paracloeodes* has claws about half the length of the tarsi while the claws and tarsi are about equal length in *Apobaetis*.

5b Gills with well-developed lateral branches on the trachea. 6

6a Wing pads present on the metathorax. .... *Centroptilum* (in part)

See also 2a. This is a widespread and common group with about 2 dozen mainly stream inhabiting species.

6b Wing pads absent on the metathorax . . . former *Neocloeon*, = *Cloeon*

A single species was once placed in *Neocloeon*, known from North Carolina and Tennessee. It is now considered to be part of *Cloeon*, a widespread genus which contains about a dozen species.

EPHEMEROPTERA ADULTS

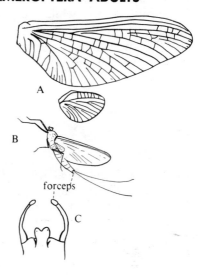

Figure 90

Figure 90 General features of mayflies. A. wings; B. adult at rest; C. male forceps

1a Appearing to have but 4 or 5 major longitudinal veins in the front wing (Fig. 91). .... Oligoneuriidae

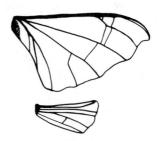

Figure 91

Figure 91 Wings of *Lachlania saskatchewanensis*

Two genera of this largely tropical family are found in America north of Mexico. *Lachlania* spp. range from New Mexico and Utah to Saskatchewan *Homoeoneuria* spp. are known from Florida to Texas, north to Nebraska and east to Indiana. The former genus has two caudal filaments (cerci) while the latter has three (2 lateral cerci plus the median terminal filament). They are strong flyers with limited periods of emergence and are rare in collections.

1b **With numerous longitudinal veins in the front wing (Figs. 94, 95).** 2

2a **Hind wing absent, front wing with anal area enlarged and rounded (Fig. 94, arrow); body size small, 4-6 mm.** 3

2b **Hind wing present but may be tiny; front wing with anal area not inflated (Fig. 96); body size variable.** 5

3a **Front wing with single or double marginal intercalary veins (arrow) (Fig. 92); lateral ocelli 1/4 or less the size of the compound eyes.** .... (in part) Baetidae

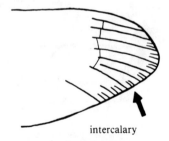

intercalary

Figure 92

Figure 92   Marginal intercalaries (arrow) of Baetidae

Apobaetis, Baetodes, Pseudocloeon, Cloeon, and *Paracloeodes* of the family Baetidae key here. Many members of this family have hind wings much reduced in size, but in the above genera the hind wings are absent.

3b **Front wings without marginal intercalaries (Fig. 93); lateral ocelli nearly half as large as the compound eyes.** 4

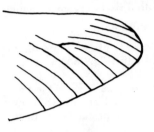

Figure 93

Figure 93   Wing lacking marginal intercalaries

4a **Crossveins few, about 10, usually in a single arc-like series across the basal part of the wing; veins R_4 and R_5 not forming a symmetrical fork (Fig. 94); forceps one segmented.** .... Caenidae

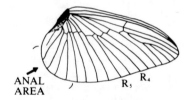

ANAL AREA R_5 R_4

Figure 94

Figure 94   Wing of Caenidae (*Brachycercus*) (anal area between brackets)

This family is considered to be most closely related to Neoephemeridae but is superficially similar to Tricorythidae. In North America it contains the two genera, *Caenis* and *Brachycercus*. Adult emergence of *Caenis* is often synchronized so that large numbers will be found fluttering over ponds or prairie potholes which are a major nymphal habitat. Adults of *Brachycercus* are less often seen. Adults of *Caenis* are able to shed the subimaginal exuviae while in flight. This occurs almost immediately after emerging from the nymphal stage. Both genera are widespread in North America and *Caenis* is very common.

4b　Crossveins more numerous, 15 or more, scattered throughout the wing; veins R_4 and R_5 forming a more or less symmetrical fork (Fig. 95); male forceps 2 or 3 segmented. .... Tricorythidae

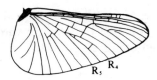

Figure 95

Figure 95　Wing of Tricorythidae (*Tricorythodes*)

Adults of this family superficially resemble Caenidae but are actually relatives of the Ephemerellidae. In both of the latter the forceps of the male are 2 or 3 segmented. Like *Caenis*, adult *Tricorythidae* can shed the subimaginal cuticle in flight. The genus *Tricorythodes* is widespread and common in North America. *Leptohyphes* is known from a few localities in the southern U.S., e.g. Texas and South Carolina, and also Maryland.

5a　Hind wing circular or oval and with numerous longitudinal veins (Fig. 96B); a bilobed structure present on the prosternum between the bases of the fore coxae (Fig. 96A); mesonotal scutellum extends posteriorly as a spinous process .... **Baetiscidae**

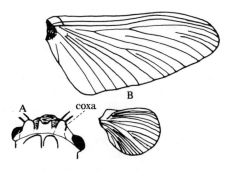

Figure 96

Figure 96　A. Ventral view of head and prothorax of *Baetisca* with bilobed process between coxae; B. wings of *Baetisca*, cross-veins omitted

Adults are easily recognized by the robustness of the body, the prosternal projection between the fore-coxae and the peculiarities of the wings. *Baetisca*, the only genus, contains about 12 species and is found primarily in eastern N.A. Several species are found in the west to Saskatchewan, Alberta, Wyoming and Washington. These bizarre mayflies are common in the nymphal stage in favorable situations, but adults are seldom collected.

5b　Hind wing more angular (Figs. 97, 100, 101); bilobed and spinous processes absent from prosternum and mesoscutellum. **6**

6a　Veins of the costal area of front wing continue around wing tip (arrow); costal crossveins absent (bracket) (Fig. 97). .... Campsurinae, Polymitarcyidae

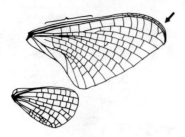

Figure 97

Figure 97 Wings of Campsurinae (*Tortopus*)

There are many species in tropical America, but in N.A. this subfamily consists of two genera, *Tortopus* with 3 species and *Campsurus* with one species. As in Ephoroninae, the middle and hind legs of adults are atrophied and functionless, but the subfamilies differ as is indicated by the key. *Campsurus decoloratus* is found in Texas; *Tortopus* spp. are found in southeast and south central U.S. and north to Manitoba.

6b   Costal veins do not continue around wing tip (Fig. 101); costal crossveins present or absent. 7

7a   Pale white mayflies about 3/4 inch in length; all legs of female and middle and hind legs of male greatly reduced and atrophied; crossveins in costal area extremely numerous. .... Ephoroninae, Polymitarcyidae

This subfamily consists of the single genus *Ephoron*, with two N.A. species, *E. leucon* and *E. album*. They are medium sized, snowy white insects which often emerge in massive numbers on river or lake shores. Females do not undergo the final moult, but mate and lay eggs as a subimago. Males have 2 "tails" while the females have 3. The genus is widespread in North America; *E. album* is a midwestern and western species while *E. leucon* is eastern.

7b   Not pale white in color nor with reduced legs; marginal veinlets present but relatively few in number. 8

8a   Vein M_2 curved and with a sharp bend near base toward Cu_1 (Fig. 98). 9

8b   Vein M_2 not bent or curved near base, M_2 more or less straight throughout its length (Fig. 99, 107, 113). 11

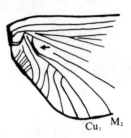

Figure 98

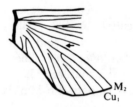

Figure 99

9a   Costal crossveins present and distinct in basal area of the front wing; costal projection of hind wing rounded (arrow) (Fig. 100). 10

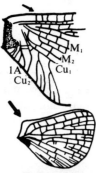

Figure 100

9b Basal costal crossveins in front wing lacking or indistinct; costal projection on hind wing sharp (Fig. 101, arrows); vein 1A unforked. .... **Neoephemeridae**

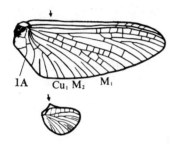

Figure 101

Figure 101 Wings of Neoephemeridae (*Neoephemera*)

In North America this family is represented by the single genus *Neoephemera* (formerly *Oreianthus*) which contains less than a half dozen species. They are found from Quebec to Michigan and south to Florida. They are unusual because the adults resemble Ephemeridae while the nymphs are similar to Caenidae. Thus the genus is placed in a separate family. These are medium sized mayflies (about 15 mm) with 3 well developed caudal filaments in both sexes and wing characters as in Fig. 101.

10a Vein 1A of front wing forked (Figs. 100, 102, arrow). .... **Potamanthidae**

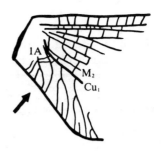

Figure 102

Figure 102 Wings of Potamanthidae

There is a single nearctic genus, *Potamanthus*, with eight species which occur in the eastern 1/2 of N.A. They are medium sized (7-13 mm) whitish mayflies with a brown thorax. Both sexes have 3 "tails". In contrast to Polymitarcidae, the legs are well developed and functional.

10b Vein 1A not forked (Fig. 103). .... **Ephemeridae**

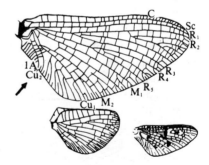

Figure 103

Figure 103 Wings of Ephemeridae

This family is made up of 4 genera and numerous species widely distributed in North America, but is most common in the east half.

Hexagenia spp. are among the largest mayflies (up to 30 mm body length, over 8 cm if tails are included). *Hexagenia* and *Ephemera* often constitute the mass emergences from the Great Lakes and Mississippi River region where drifts of mayfly bodies may be several feet deep on roads and bridges and snowplows may be required to move them. The number of caudal filaments varies with the sex and genus. Wings often have a distinct pigmented pattern (Fig. 103). *Litobrancha* was recently described by McCafferty (1975) to include one eastern species formerly in the genus *Hexagenia; Pentagenia,* while not rare, is infrequently collected. It is sometimes placed in its own family, Palingeniidae.

11a Outer half of front wing lacks forks in veins (Figs. 104, 105). 12

11b One or more forks in veins in outer half of front wing (Fig. 107, veins R₄ and R₅). 13

12a Anal area of front wing angular (arrow); marginal intercalaries absent; body 13-16 mm long (Fig. 104). .... **Behningiidae**

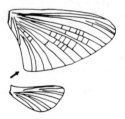

Figure 104

Figure 104   Wings of Behningiidae (*Dolania*)

These are rare and unusual insects with dorso-ventrally flattened white and purplish-brown bodies which occur from South Carolina to Florida. In North America, the only species in the family is *Dolania americana.* The penes of

the male are longer than the forceps. A similar genus, *Behningia,* is found in Russia and Poland.

¹2b Anal area of front wing rounded (Fig. 105); marginal intercalaries present (Fig. 92); body under 12 mm in length. .... *Baetidae*

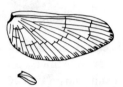

Figure 105

Figure 105   Wings of Baetidae (*Baetis*)

This widespread and common family consists of about 10 North American genera and over 100 species. The family is characterized by reduced wing venation, the reduction in size or loss of the hind wing, and the three segmented middle and hind tarsi. Many members of this family are tiny, 4 mm or less, but *Callibaetis* and *Centroptilum* may be 10 mm or more. While this family is easy to recognize, identification at the genus and species level is very difficult because of the lack of structural characters of taxonomic value. Even though common, this remains one of the most inadequately studied groups of mayflies.

13a With two pairs of longitudinal veins (cubital intercalaries) at hind angle of front wing between Cu₁ and Cu₂ (Figs. 106-109). 14

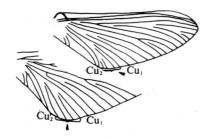

Figure 106

Figure 106 Wings of *Ametropus* (upper) and *Epeorus* (lower)

13b Without two distinct pairs of veins at hind angle of front wing (Figs. 110-113). 17

14a Five tarsal segments in the hind leg. .... **Heptageniidae**

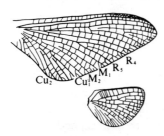

Figure 107

Figure 107 Wing of Heptageniidae (*Epeorus*)

While nymphs of Heptageniidae are separated to subfamily and genus in this key, all subfamilies of adults (Heptageniinae, Anepeorinae, Arthropleoninae) except Pseudironinae (with 4 tarsal segments) key to this point. The adults of Spinadinae are unknown. Adults are not easily separated to genus. This family is readily recognized (except Pseudironinae) by the 5 tarsal segments in the rear leg plus the two pairs of cubital intercalaries in the front wing. The group is widespread, common, and contains about 10 genera and over 150 species.

14b With 4 tarsal segments in the hind legs. 15

15a With three caudal filaments (wing, Fig. 106). .... **Ametropodidae**

Ametropodidae, as now used, contains only the genus *Ametropus*. The family formerly encompassed *Pseudiron* (now Heptageniidae) and the genera now included in Acanthametropodinae (Siphlonuridae), and the Metretopodidae. This situation arose both because of the rarity of some species and thus the lack of material for study and the fact that often nymph and adult characters are contradictory regarding family affinities.

Ametropus, according to recent studies, contains 3 species which range from Saskatchewan and Alberta through Montana and Oregon to Utah and New Mexico. While generally very rare, they are locally abundant.

15b With two caudal filaments. 16

16a Wings clear, without dark markings; first tarsal segment of the male front leg about 1/2 to 3/4 as long as the second (Fig. 108); female with subanal plate notched at the midline. .... **Pseudironiinae, Heptageniidae**

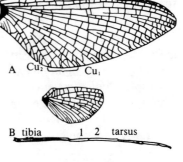

Figure 108

Figure 108 *Pseudiron.* A. Wings; B. Male front leg

Because the adult wing venation is typical of the Heptageniidae with two pairs of cubital intercalaries, this subfamily is now considered to be part of Heptageniidae rather than Ametropodidae as is found in older books. It consists only of the rather rare genus *Pseudiron* with 2 species. *P. centralis* ranges from Saskatchewan and Manitoba to Kansas and Missouri. *P. meridionalis* occurs in Georgia and Florida.

16b **Hind wing almost always clouded with brown at the base, front wing with brown speckles in the costal region; first tarsal segment of the male front leg usually as long as the second; females with subanal plate not notched at the midline (Fig. 109).** .... Metretopodidae, in part, *Siphloplecton*

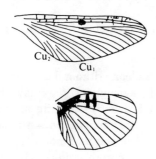

Figure 109

Figure 109 Wings of *Siphloplecton,* hind wing enlarged, front wing without crossveins

The family Metretopodidae consists of the genera *Metretopus* and *Siphloplecton* (see also Couplet 18b). *Siphloplecton* contains about half a dozen species which range from Florida to Quebec and west to Saskatchewan and Minnesota. It is fairly common.

17a **With two caudal filaments.** 18

17b **With three caudal filaments.** 19

18a **Numerous veins attached to the inner hind angle of the wing between Cu$_1$ and Cu$_2$ (Fig. 110).** .... Siphlonuridae

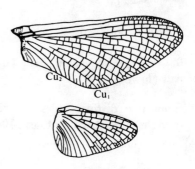

Figure 110

Figure 110 Wings of Siphlonuridae (*Siphlonurus*), some crossveins omitted

The family Siphlonuridae in North America consists of three subfamilies, two of which (Siphlonurinae and Isonychiinae) key to this point. The third subfamily, Acanthametropodinae, is found at Couplet 19a. Siphlonurinae adults have cubital intercalaries which form a series of unbranched parallel veins attached to the wing margin (Fig. 110) and there are no gill remnants at the bases of the fore-coxa. In Isonychiinae, with the single genus *Isonychia,* the vein pattern is similar except that the cubital intercalaries are sinuate and branched, and there are gill remnants on the maxillae and coxal bases. The family is common and widespread in North America. *Siphlonurus* and *Ameletus* are common genera.

18b **With a single pair of veins between Cu$_1$ and Cu$_2$ (Fig. 111).** ....
....Metretopodidae, in part, *Metretopus*

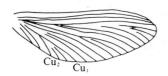

Figure 111

Figure 111 Wing of *Metretopus*, crossveins omitted

Metretopus contains the single species *M. borealis* which is holarctic, being found in Scandinavia as well as several Canadian provinces (B.C., Alberta, Saskatchewan, Manitoba, and N.B.), and Maine and Michigan. See also Couplet 16b, *Siphloplecton*.

19a Vein Cu_2 not strongly curved; with numerous intercalaries between Cu_1 and Cu_2 (Fig. 112). ....
.... **Acanthemetorpodinae, Siphlonuridae**

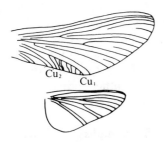

Figure 112

Figure 112 Wings of *Analetris eximia*, crossveins omitted

This is a poorly known and rare group (but see Couplet 18a). *Siphluriscus* (from China) is known from the adult stage only. *Acanthametorpus* (= *Metreturus*) is known only from the nymphal stage and is found in Illinois, Georgia, South Carolina, as well as Siberia. *Analetris eximia,* known in both the nymph and adult stage, is found from Saskatchewan to Wyo-

ming and Utah. Note the large size of the hind wing (Fig. 112).

19b Vein Cu_2 strongly curved (Fig. 113); cubital intercalaries as in Figs. 113 or 114. .... **20**

20a Male forceps with one short terminal segment; with 1 or 2 long intercalary veins between vein Cu_1 and M_2 (Fig. 113). ......
.... **Ephemerellidae**

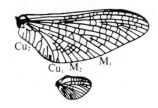

Figure 113

Figure 113 Wings of Ephemerellidae (*Ephemerella*)

This family contains only the genus *Ephemerella* which is divided into 8 North American subgenera (see nymphal key). Members are common and widely distributed. The 3 tails and characteristic wing venation make this group easily recognizable. Adults are strong fliers and are generally not abundant in collections.

20b Male forceps with two short terminal segments; without long intercalary veins between Cu_1 and M_2. .. **Leptophlebiidae**

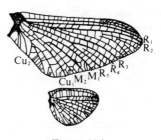

Figure 114

Figure 114 Wings of Leptophlebiidae (*Leptophlebia*)

This common and widespread family contains 7 North American genera (see nymphal key) and over 50 species. The family is easily recognized by the three tails plus the characteristic wing venation. Adults are rather delicate and medium sized and often have a reddish color.

ODONATA
Dragonflies and Damselflies

The term dragonfly formerly referred to all members of the Order Odonata but current usage restricts "Dragonfly" to the Anisoptera and "Damselfly" to Zygoptera.

Odonata are predatory in both adult and nymphal stages. (Many use the term "naiad" rather than "nymph" in this Order.) Adults (Figs. 37, 128) are voracious predators which seek their prey visually, attacking flying or resting insects ranging from tiny Diptera to other Odonata of considerable size. Victims are caught up in basket-like, spiny and anteriorly directed legs characteristic of the Order and are then devoured with the strongly developed mouthparts.

Males have forcep-like appendages on the end of their abdomen used for grasping the head (dragonflies) or thorax (damselflies) of the female during mating. Prior to mating sperm of the male is transferred from an opening in the ninth abdominal segment to a structure under the second abdominal segment. During mating, the female, held by the forceps of the male, reaches forward with her abdomen and receives the male sperm with the genital opening on the eighth segment. In many species of Odonata the female has a well developed spine-like ovipositor at the genital opening. Eggs are deposited in living or dead plant material. Details of mating and egg laying behavior vary considerably among the species.

Nymphs are all predatory, capturing prey from tiny Crustacea to various insects and other invertebrates to small fish and tadpoles (Fig. 115). The mouthparts are remarkable in that the labium (Figs. 18A, F; Fig. 120), ending in a pair of spiked jaws, can be extended or retracted by means of an elbow-like articulation (Fig. 18A). Once prey is captured the victim is drawn to the mandibles where it is chewed up and devoured. Damselfly nymphs respire by means of leaf-like gills at the end of the abdomen (Fig. 115C, D). Most of this group climb plants or scramble over trashy substrates. Dragonfly nymphs range from burrowers to plant climbers and respire by means of a rectal chamber which pumps water in and out. They are often seen to be "jet propelled" in a collecting pan because of the stream of water they are able to squirt from the anus. Some odonate nymphs lie in ambush while other individuals stalk and attack their prey.

Figure 115 Nymphs of Odonata. A. Aeshnidae; B. Gomphidae; C. Coenagrionidae; D. Gill of Coenagrionidae; E. Libellulidae

Because of their size and color, Odonata have long been the delight of entomological collectors. Unfortunately the brilliant reds, blues and greens of damselflies and dragonflies usually fade in the preserved state. Nymphs are traditionally preserved in 70 or 80% ethyl alcohol, but I personally have found a small pinned collection to be very useful. Adults should be pinned in the usual way, or they may be dried flat and stored in clear coin envelopes in a filing cabinet.

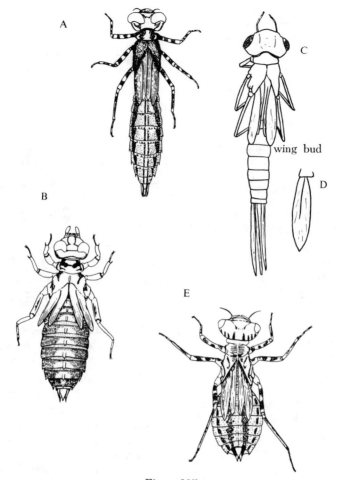

Figure 115

ODONATA NYMPHS*

1a Abdomen terminates in sharp spines (Fig. 116) (Dragonflies). .... **Anisoptera, 4**

Figure 116 Terminal spines of Dragonfly nymphs

1b Abdomen terminates in three flat, leaf-like gills (Fig. 117) (Damselflies). .... **Zygoptera, 2**

*Protoneuridae, a neotropical family of stream damselflies represented in Texas by two species, is not included in this key; see Johnson (1972).

Gills of nymphs are thickened at the base and are distinctly thinner in the outer half; in adults the quadrangle (Fig. 136) is more triangular than trapezoidal.

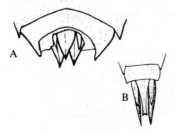

Figure 116

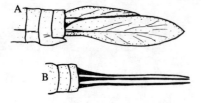

Figure 117

Figure 117 Lateral (A) and dorsal (B) view of gills of Damselflies

2a **Individual segments of the antenna about equal in length (Fig. 118). 3**

Figure 118

Figure 118 Antenna

2b **First segment of antenna as long as following 6 segments combined (Fig. 119) (Stream Damselflies). .... Agrionidae (= Agriidae = Calopterygidae)**

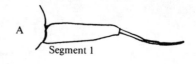

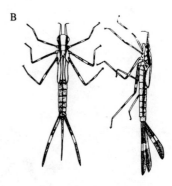

Figure 119

Figure 119 A. Antenna of Agrionidae; B. Dorsal and lateral views of Agrionidae

These nymphs are stream inhabitants which cling to and climb over roots and trash at the edge of the current. There are two North American genera.

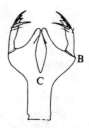

Figure 120

Figure 120 Labial mask of *Agrion*

In the genus *Agrion,* the labial mask divided about half its length by a cleft (C) which extends deeper than the base (B) of the lateral lobes (Fig. 120). These are the largest North American stream damselflies. They are awkward, long-legged insects that cling to stems and roots in the current. They move very little from place to place. While localized in distribution, they are widespread and common in North America. The mask in *Hetaerina* has a cleft extending only to the bases of the lateral lobes, and thus the mask is divided much less than half its length by the cleft. These nymphs are similar in habits and appearance to *Agrion* (Fig. 119B). They are widespread and locally abundant wherever suitable streams are found.

3a   Basal part of labial mask narrowed to a stalk when viewed from below (Fig. 121). .... **Lestidae**

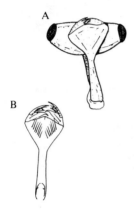

Figure 121

Figure 121   A. Ventral view of head of Lestidae; B. detail of mask, dorsal view

These are usually found in marshy areas or protected margins of lakes. They are slender climbing forms. The gills are usually parallel-sided and blunt-tipped.

3b   Basal part of labial mask not narrowed to a stalk when viewed from below, somewhat triangular in shape (Fig. 122). .... **Coenagrionidae**

Figure 122

Figure 122   Ventral view of head of Coenagrionidae

These nymphs are slender greenish forms which clamber over vegetation and debris in ponds and marshes. Some are found in gently moving water. Gills are usually lanceolate and pointed. These are the most common damselflies.

4a   Labial mask over mouth is spoon-like and covers face nearly to eyes (Fig. 123). **5**

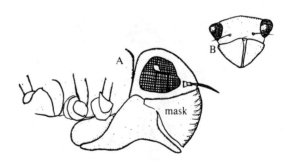

Figure 123

Figure 123   Lateral and front view of head and labial mask

4b   Labial mask over mouth not spoon-like but flat in side view (Fig. 124). **6**

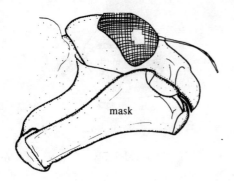

Figure 124

Figure 124   Ventro-lateral view of head and mask

5a   In face view the mid-seam of the mask even or with regular saw-like teeth (Figs. 123B, 125B). .... Libellulidae

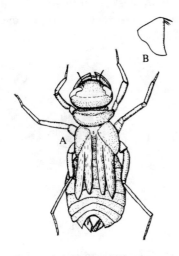

Figure 125

Figure 125   A. Libellulidae; B. front view of mask (half)

Nymphs in this family are bottom sprawlers with widely set legs which clamber over vegetation and debris (Fig. 125). They seldom burrow or climb vegetation. While subfamilies are separated in the adult key, no single set of characters will separate subfamilies in the nymphal stage. Macromiinae are recognized by a distinct erect frontal horn between the eyes. They live in large streams and lakes where there is considerable water movement. Nymphs of Libellulinae prefer warm, shallow, eutrophic water while Corduliinae prefer streams and lakes with high oxygen content. These two subfamilies are not easily separated.

**5b   In face view the mid-seam of the mask with large irregular teeth (Fig. 126). ......
.... Cordulegastridae**

Figure 126

Figure 126   Face view of Cordulegastridae

These nymphs are stout, hairy and cylindrical, tapering beyond the middle to a pointed posterior apex. They live in clear woodland streams buried in sand or silt waiting in ambush for passing prey. They are uncommon.

**6a   Antennae with 4 segments, segment 3 enlarged, segment 4 minute (Fig. 127). ......
.... Gomphidae**

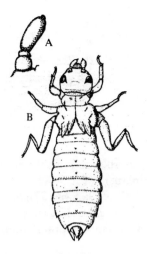

Figure 127

Figure 127 A. Antenna of Gomphidae; B. Nymph

Nymphs of this family can be immediately recognized by the characteristic form of the antennae. They usually develop in flowing water but a few species can be found in lakes or even marshy ponds where they sprawl or burrow into the substrate. Because the legs are set laterally they are unable to climb plants.

6b   Antennae with 6 or 7 segments of similar length (Fig. 118). 7

7a   Antennae slender and without black hair-like bristles; common (Fig. 115A). .... Aeshnidae

These nymphs are climbers and they live in tangles of vegetation and debris in both still and flowing water. They often actively stalk their prey.

7b   Antennal segments short and thick, with heavy black bristles covering all segments of antenna; rare. .... Petaluridae

Nymphs are very rare. They have been found under moss in seep areas near springs. See Couplet 9b, adult key.

ODONATA ADULTS*

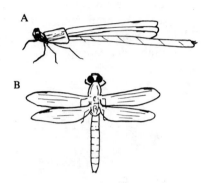

Figure 128

Figure 128   Adult Odonata. A. Zygoptera; B. Anisoptera, diagrammatic

1a   Front and hind wings similar in shape at the base, either narrow with abrupt widening (Fig. 129A) or evenly widening (Fig. 129B) (Damselflies, Suborder Zygoptera). 2

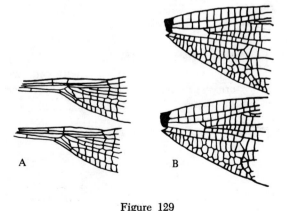

Figure 129

*See footnote, page 75.

Figure 129   Wing bases of Zygoptera

These are the smaller (usually about 3 cm) and more delicate members of the Order, usually found fluttering among grasses and reeds on the shores of lakes and ponds. Appearance is as in Fig. 37A and 128A.

1b   Front and hind wings dissimilar in shape at the base, hind wing much broader than the front wing (Fig. 130) (Dragonflies, Suborder Anisoptera). 4

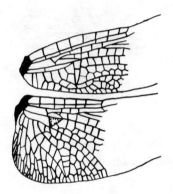

Figure 130

Figure 130   Wing base of Anisoptera

Body size ranges from about 2 to 8 cm in length. These are the strong flying, aggressive and wide ranging Odonata (Figs. 37B and 128B).

2a   Wings narrow at the base but suddenly become wider (petiolate); with but two crossveins between the nodus and base (Fig. 131). 3

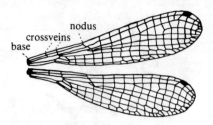

Figure 131

2b   Wings rounded evenly at the base (not petiolate) (Fig. 132); with numerous crossveins between nodus and base (Stream Damselflies). .... Agrionidae (=Agriidae=Calopterygidae)

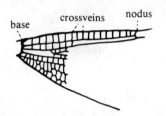

Figure 132

Figure 132   Wing base of Agrionidae

Adults have been referred to as "the birds of paradise" of the Odonata. There are but two North American genera.

Figure 133

Figure 133   Wing of *Agrion*

The Black Wings (*Callopteryx* = *Agrion*) are large metallic green damselflies. The wings are tipped with black or are entirely black (Fig. 133). These are elegant metallic damselflies which have a characteristic fluttering and wavering butterfly-like flight. There are half a dozen North American species; the genus is widespread except for the prairies and far north. The female oviposits unattended by the male. The female backs into flowing water the length of the abdomen (or even up to a foot under water) and deposits the eggs in plant stems or rotting sticks.

Figure 134

Figure 134   Wing of *Hetaerina*

The Ruby Spots (*Hetaerina*) are slender blackish or bronze dragonflies with a ruby, reddish, or amber basal wing spot (Fig. 134). This genus is mostly tropical but there are four North American species. *H. americana* is widespread and the genus can be found from coast to coast. Adults frequent riffles of small, clear streams. They seldom venture far away from the water. Oviposition is similar to that in *Agrion.*

3a   A small triangular cell in wing midway between nodus and hind edge of wing, i.e. vein M$_3$ arising nearer the nodus than the arculus (Figs. 135, 131) (Narrow-winged Damselflies). ....
.... (Coenagriidae) = Coenagrionidae

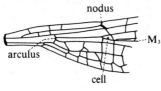

Figure 135

Figure 135   Wing of Coenagrionidae

This is a variable group containing over a dozen genera and nearly 100 North American species. At rest, the wings are held folded against the abdomen. In general these are small damselflies (body 2-4 cm) with body colors tending towards black, red and blue, and with clear wings. These are probably the most frequently seen damselflies with some genera wandering far from the marshy nymphal habitat. Females generally oviposit without males in attendance by thrusting the ovipositor as far under water as possible while resting on floating objects; sometimes eggs are laid in the stems of plants.

3b   Without a triangular cell behind nodus; i.e. vein M$_3$ arising nearer the arculus than the nodus (Fig. 136) (Spread-winged Damselflies). .... Lestidae

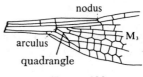

Figure 136

Figure 136   Wing of Lestidae

Because adults often hold the wings obliquely upward and backward while at rest, these are sometimes called spread-winged damselflies. There are but two North American genera. *Lestes* (about a dozen spp.) is widespread and common; *Archilestes* (2 spp.) is found from Washington to California. Members of the family are medium to large (body 3-5 cm). Colors are usually brown to bronzy green. Adults prefer the seclusion of dense reeds and marshy vegetation. During oviposition, which occurs in plant stems below or a few inches above the water, the male sometimes holds the female's thorax with his forceps while she deposits eggs in a stem with her ovipositor.

4a   Triangle of fore and hind wings about equally distant from arculus (Fig. 137).
.... 7

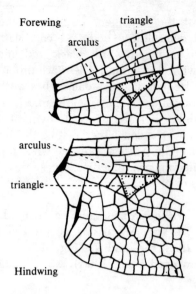

Figure 137

Figure 138 Wings of Libellulidae (Corduliinae)

These are the most common and best known dragonflies, very showy and often seen hovering over ponds. Wings are often pigmented. The eyes meet at the top of the head in a short seam (Fig. 143) and the triangles of the wings are oriented in different directions: crossways in the front wing and lengthways in the hind wing. Body color is often beautiful—red, gold, brown, sometimes metallic, and in old age sometimes the abdomen is covered with white powder. The body is stout and less elongate than in most other groups. Three rather distinct subfamilies are recognized and are separated in the following key.

4b **Triangle more distant from arculus in fore wing than hind wing (Fig. 138), females generally lack an ovipositor.** .... **Libellulidae, 5**

5a **Hind wing with triangle some distance from arculus (Fig. 139). .. Macromiinae**

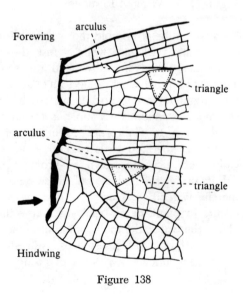

Figure 138

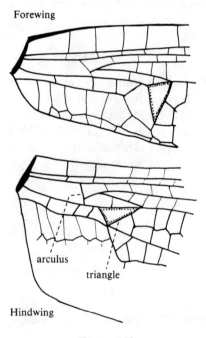

Figure 139

Figure 139   Wings of Macromiinae

This is a small group with two North American genera (*Didymops* and *Macromia*) and about a dozen species. The group has a wide distribution but is absent from the prairies and far north. They are strong, high, wide ranging fliers which are difficult to capture. The body is large and *Aeshna*-like, and an oblique yellow band belts the thorax between the clear wings. Females fly over the water touching the surface periodically to release eggs.

5b   **Hind wing with triangle close to arculus (Figs. 138, 140).** 6

6a   **Anal loop (stippled, Fig. 140) generally distinctly foot-shaped; inner rear angle of hind wing rounded (arrow); males without an ear-like lobe on abdominal segment 2.** .... **Libellulinae**

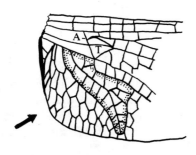

Figure 140

Figure 140   Wing of Libellulinae (A—arculus; T-triangle)

This group is large, containing many North American genera of Dragonflies. It contains many of the frequently seen medium-sized dragonflies and is abundant in warm regions. There are few species in the far north. Colors are often striking in this group, but they are not metallic. Females fly over the water and dip the abdomen to the surface to release eggs.

6b   **Anal loop not foot-shaped, or with no development of toe; in males, inner rear angle of hind wing with a triangle of cells and a distinct notch on the inner margin (Fig. 138); males with a small ear-like lobe on the side of abdominal segment 2.** .... **Corduliinae**

Like the previous subfamily, this is a large group of medium-sized dragonflies with about a dozen genera and numerous species, many of which are commonly seen. Coloration is often dark with a metallic or brassy luster. The group has a predominantly northern distribution and many are found in the northern forests of Canada. Females fly over the water and release eggs on the water surface.

7a   **Stigma with a crossvein at inner end (Fig. 141, arrow).** 8

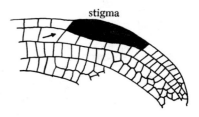

stigma

Figure 141

7b   **Stigma without a crossvein at inner end (Fig. 142).** .... **Cordulegastridae**

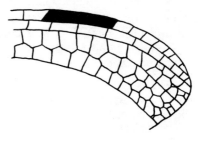

Figure 142

Represented in North America by a single genus, *Cordulegaster*, it contains eight species and can be found throughout the continent where there are suitable nymphal habitats. They are large (body 5-7 cm), clear winged, yellow and black insects with the eyes touching at a single point on the top of the head. Adults patrol clear woodland streams, and in the west frequent mountain gorges and the vicinity of rushing torrents. Adults are difficult to capture. Females have been observed in oviposition hovering perpendicularly over shallow water and stabbing the abdomen into sand.

8a   **Eyes meeting in a long seam on the top of the head (Fig. 143). .... Aeshnidae**

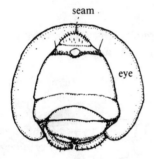

Figure 143

Figure 143   Face view of head, Aeshnidae

These are large (body 5-7 cm long) clear winged, strong fliers which venture far from water and often dominate the air over forest clearings, meadows or ponds. The group contains a number of genera and species; the color pattern is brownish, spotted with yellow, blue or green. Segments 1 and 2 of the abdomen are inflated. The female has a well-developed ovipositor. Unaccompanied by a male, or sometimes in tandem, she lays her eggs in soft stems a little below the water surface.

8b   **Eyes widely separated on the top of the head (Fig. 144). 9**

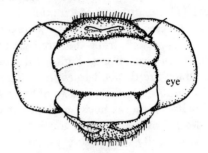

Figure 144

9a   **Stigma relatively short and wide and with a bulge near the middle; labium not notched at midline (Fig. 145). .... Gomphidae**

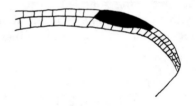

Figure 145

Numerous genera and species make up this group of medium to large dragonflies. Adults have clear wings but the bodies are often brightly patterned with black, green and yellow. Adults characteristically perch on low objects or even on the ground from where they take short flights. They do not "patrol" as do some other groups. They prefer sunshine and are not active on dull days or in the evenings. During oviposition, females are unattended by males. They fly slowly over the surface of the water, descending irregularly to release eggs.

9b   **Stigma long, narrow and parallel-sided, no bulge at mid-length; labium deeply notched at midline (Fig. 146). .... Petaluridae**

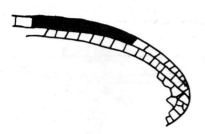

Figure 146

Adults of these uncommon dragonflies are large (body length 5-7 cm) hairy, rather grayish in color and with the eyes widely separated at the top of the head. Nymphs are thick set, hairy, and rough in appearance. They live in seeps and bogs. Eggs are placed among roots and vegetation above the water surface by un-accompanied females. There are two N.A. genera represented by a single species each, *Tanypteryx hageni* (the Western Grayback from B.C., Nevada, Washington, Oregon, California) and *Tachopteryx thoreyi* (the Eastern Grayback from Florida to New York and Michigan to Texas). Adults of the Western Grayback have been collected in alpine meadows and nymphs (very rarely collected) have been found in a blanket of wet moss where a trickle flowed over a solid rock substrate. Adults of the eastern species are often seen resting on tree trunks and rail fences near the nymphal habitats, which are small springs or meandering streams which course through boggy, wooded valleys.

PLECOPTERA
Stoneflies

Stoneflies are a rather uniform group of insects both in structure (Figs. 148, 155) and behavior. Nymphs are restricted to flowing water or standing water which is cool and high in oxygen content. The Suborder Filipalpia, characterized by having glossae and paraglossae (Figs. 10, 150A) of about equal length, consume plant material in the nymphal stage, with very few exceptions. Filipalpia includes all Families except Perlidae, Perlodidae, and Chloroperlidae. The latter belong to the Suborder Setipalpia with paraglossae longer than the glossae (Fig. 150B) and are carnivorous, feeding on other insects and any stream animals they can catch. Recent systematic studies no longer use the above suborder classification. Stonefly nymphs are important members of aquatic ecosystems and are frequently very abundant in both kinds and members. They are often major members of the food chain leading to game fish.

The majority probably do not feed in the adult stage, but records show that some species feed on algae, willow catkins, and plant buds, and also drink water.

Adults are clumsy flyers that do not venture far from the nymphal habitat. In some species wings of adults are reduced to 1/2 or 1/4 the length of the abdomen (brachyptery).

All stages should be collected and preserved in 70 or 80% alcohol.

PLECOPTERA NYMPHS

1a   With bushy gills both on the ventral thorax *and* on the first two or three segments of the abdomen (Fig. 147). ....
.... **Pteronarcidae**

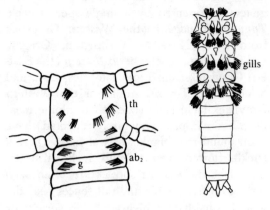

Figure 147

Figure 147 Ventral gills (g) of Pteronarcidae, highly diagrammatic; th, thorax; Ab,
abdomen

These are large (3 cm or more when mature)
primitive stoneflies with gills ventrally on the
anterior abdominal segments. Glossae and
paraglossae are of equal length. Nymphs feed
on leaves or other organic material and development takes two years or more. They are
found in flowing water of all types.

1b **Without bushy gills on the abdomen, but
gills may be present on the ventral tho-
rax; size variable.** 2

2a **With bushy gills on the ventral thorax at
the bases of the legs (coxae) (Fig.
148B).** .... Perlidae

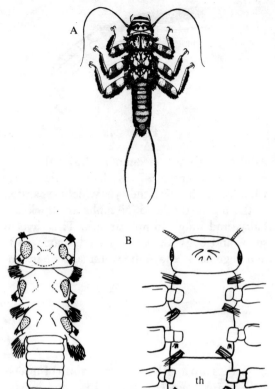

Figure 148

Figure 148 A. Perlidae; B. Ventral gills (g),
highly diagrammatic; th, thorax; Ab, abdomen

Glossae are much shorter than the paraglossae
(Fig. 150B). Nymphs of this group are predatory and feed on any stream animals they can
catch, especially other aquatic insects. They
generally prefer cool clear streams and may
take up to 3 years to develop. Many species are
well over 3 cm in length when mature, but others are less than 2 cm.

2b **With only small, finger-like gills or none
at all.** 3

3a Prothorax (P) at least 1 1/2 times the width of the head (Fig. 149B); body broad and roachlike (Fig. 149A); two ocelli (o) .... Peltoperlidae

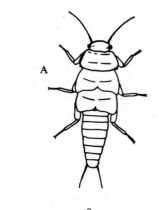

Figure 149

Figure 149   A. Peltoperlidae; B. detail showing ocelli (o) and prothorax (p)

Glossae and paraglossae are equal in length (Fig. 150A). Nymphs feed on decaying leaves and detritus in the cool streams in which they live. They are medium sized (10-15 mm) and probably require 1 year to develop.

3b Prothorax about equal in width to the head; body elongate, ocelli usually 3 (Figs. 151, 152). 4

4a Glossa and paraglossa about equal in length (Figs. 10, 150A). 6

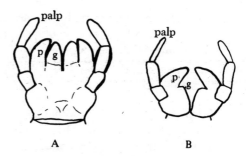

Figure 150

Figure 150   Labium with glossa (g) and paraglossa (p). A. Filipalpia, B. Setipalpia

4b Paraglossa much longer than glossa (Fig. 150B). 5

5a Cerci (C) about equal in length to the abdomen, or longer; gills usually absent but are fingerlike when present; body often with a distinct color pattern. .... Perlodidae

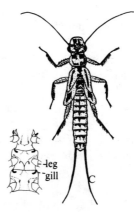

Figure 151

Figure 151   Perlodidae

These are common in cool clear streams, with some species inhabiting warm prairie rivers. In mature nymphs the hind wing pads diverge at an angle from the long axis of the body. They are medium in size (10-20 mm when mature); the glossae and paraglossae are un-equal in length, and they feed on other insects and stream animals. Development time of nymphs is usually 1 year. *Isoperla* is the most common genus.

5b Cerci (c) about 1/2 (sometimes 2/3) as long as the abdomen (Fig. 152); gills absent; body uniform in color. ....
.... **Chloroperlidae**

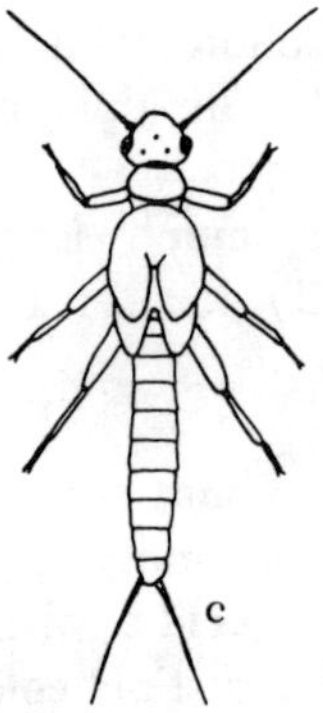

Figure 152

Figure 152 Chloroperlidae

These predatory nymphs are common across North America in clear flowing water. In mature nymphs the wing pads are parallel to the axis of the body. They are medium sized and usually require 1 year to develop. *Alloperla,* the most common genus, has recently been split into a number of genera.

6a Second tarsal segment about as long as the first (Fig. 160A). ..... **Taeniopterigidae**

Nymphs of these small to medium sized stone-flies feed on decaying leaves, algae, and detritus. Some are known to enter an inactive stage in summer. Growth and development take place in winter and adults emerge in late winter or early spring while snow and ice are still present. Until recently all members were con-

sidered to belong to the genera *Brachyptera* and *Taeniopteryx* but these have now been split into 6 genera. Cerci are long in nymphs but short in adults. Appearance is as Fig. 153.

6b Second tarsal segment distinctly shorter than the first (Fig. 162B). 7

7a Wing pads of mature nymphs bent away from the body axis (Fig. 153); body stout. .... **Nemouridae**

Figure 153

Figure 153 Divergent wing buds (Taeniop-terigidae and Nemouridae)

These are small stoneflies which feed on organic material and which prefer small streams or springs and seep areas. Until recently all North American species were placed in the genus *Nemoura* but 12 genera are now recognized. Cerci are long in nymphs but short in adults.

7b Wing pads of mature nymphs not bent away from the body axis (Fig. 152); nymphs more elongate and cylindrical. 8

8a In side view, abdominal segments 1-9 consist of an upper and lower portion, each segment divided on the side by a membranous fold (compare with Fig. 154). (The use of any standard biological stain or dye will often make the fold easier to see.) .... Capniidae

These are small, detritus feeding stoneflies that live in a variety of small streams. There are 9 genera; *Allocapnia* is the common eastern genus while *Capnia* is dominant in the west. Cerci are multi-segmented in both nymphs and adults. The abdomen is often flat ventrally.

8b   At most, segments 1-7 divided laterally by a membranous fold (Fig. 154). .... **Leuctridae**

Figure 154

Figure 154   Abdomen, lateral view, Leuctridae

These stoneflies favor the smaller and slower streams and feed on dead leaves and detritus. There are 6 genera and the group is widespread and fairly common. The posterior segments of the abdomen are circular in cross-section and form a solid ring.

PLECOPTERA ADULTS

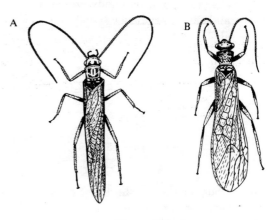

A   B

Figure 155

Figure 155   Adult Stoneflies

1a   Anal area (bracket, Fig. 156) of front wing with a network of crossveins; abdominal segments 1, 2 and sometimes 3 with remnants of larval gills (compare with Fig. 147). .... **Pteronarcidae**

Figure 156

Figure 156   Front wing, Pteronarcidae

This primitive family is common and widespread and contains about 10 species divided into three genera (*Allonarcys, Pteronarcella* and *Pteronarcys*). They are up to five cm in length from the head to the tip of the folded wings. Smaller species are about 2 cm in length. Adults have multi-segmented cerci and numerous crossveins in the wings; glossae and paraglossae are subequal in length. Adults do not feed but often drink water. Adults may live nearly a month and lay 1000 eggs.

1b Crossveins absent in anal area (bracket, Fig. 157); gill remnants not present on abdominal segments 1 and 2; size variable. 2

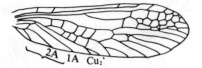

Figure 157

2a Abdominal cerci short, not more than 10 segments in length and not longer than the width of the pronotum (Fig. 158). ....
.... 3

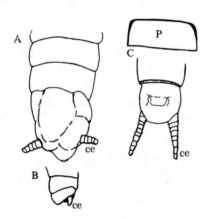

Figure 158

Figure 158 A. Abdomen of Taeniopterygidae (ventral); B. Nemouridae (lateral); C. Pronotum and abdomen (ventral) of Peltoperlidae; ce—cerci, p—pronotum

2b Cerci long, distinctly longer than the width of the pronotum. 7

3a Body relatively short and broad, somewhat roachlike (compare with Fig. 149A); with two ocelli; 10 or more costal crossveins in the front wing (bracket, Fig. 159). .... Peltoperlidae

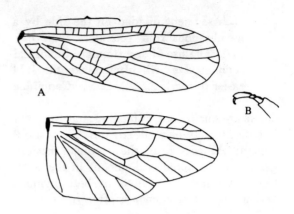

Figure 159

Figure 159 *Peltoperla*, A. wings, B. tarsus

This family was formerly considered to consist of the single genus *Peltoperla*, but 5 genera and 13 species are now recognized. These are common in the mountainous eastern and western regions of North America, but are uncommon in the interior even when habitat appears suitable.

3b Body form elongate (compare with Fig. 155); fewer than 10 costal crossveins in the front wing; with two or three ocelli.
.... 4

4a Second tarsal segment about equal to the first in length (Fig. 160A). ....
.... Taeniopterigidae

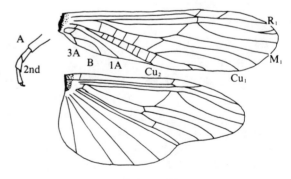

Figure 160

Figure 160 Taeniopterygidae, A. tarsus, B. wings

These are medium sized (8-15 mm) usually dark stoneflies which are widespread and common. They are often called winter stoneflies because most emerge when snow and ice are present. The glossae and paraglossae are about equal in length as are the three tarsal segments. Cerci are single or several segmented. There are numerous crossveins between Cu and M (Fig. 160B). The family has been divided into 6 genera and contains about 30 species.

4b First and third tarsal segments much longer than the second segment (Fig. 162B). **5**

5a Cerci long, with at least 4 segments; only one or two crossveins between Cu and M. .... some Capniidae

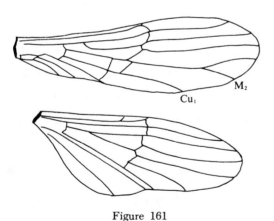

Figure 161

Figure 161 Capniidae, wings

Capniidae (small winter stoneflies) are 5-10 mm long, dark-colored insects which are widespread and common wherever there are clear cool streams. The family contains about 130 species and has been divided into 9 genera.

The major genera are *Allocapnia*, mainly eastern, and *Capnia*, mainly western. Many species are among the winter stoneflies, i.e. emerge onto snow and ice at freezing or near freezing temperatures. Cerci are multi-segmented and somewhat variable in length. There are but 1 or 2 cubital crossveins in the front wing. The glossae and paraglossae are equal in length.

5b Cerci of one segment only; 4 or more crossveins between Cu and M (Fig. 162). **6**

6a Apical crossvein (apc) present (Fig. 162); wings held flat over body when at rest (Fig. 155B). .... Nemouridae

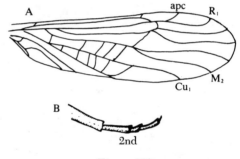

Figure 162

Figure 162 Nemouridae, A. wing, B. tarsus; apc, apical crossvein

Nemouridae formerly consisted of the single genus *Nemoura* but recent work proposes that it be divided into twelve genera in North America. All are small (5-12 mm) dark colored stoneflies; cerci are 1 segmented, glossae and paraglossae are equal in length, wings have numerous Cu and M crossveins, and the wings lie flat over the back when at rest. There is also a slanting crossvein (apc) connecting the costa and R_1 of the front wing. The family is widely distributed in North America wherever there are cool streams or springs and it contains about 60 species. Emergence is usually in the spring of the year.

6b Apical crossvein absent (Fig. 163); wings rolled around body when at rest (Fig. 155A). .... **Leuctridae**

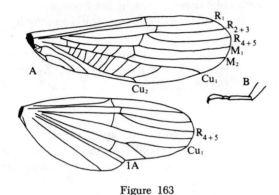

Figure 163

Figure 163   Leuctridae, A. wings, B. tarsus

Leuctridae consists of 6 genera and about 40 N.A. species. These are small (6-10 mm) and dark in color. There are numerous Cu and M crossveins and cerci are single segmented. Glossae and paraglossae are equal in length and the middle tarsal segment is shorter than the first and third. Adults can be found from spring until autumn and are widespread and common.

7a First and third tarsal segments about equal in length and 1 or 2 Cu and M crossveins in front wing (Fig. 161). .... some Capniidae, return to Couplet 5a

7b Third tarsal segment much longer than the first; several to many Cu and M crossveins. **8**

8a Remnants of branched and tufted nymphal gills present on the ventral thorax at the base of the coxae (Fig. 148B). .... **Perlidae**

This family is made up of about a dozen genera and three dozen species in North America. *Acroneuria* is the most common genus. Length in this family ranges from less than 10 to over 40 mm. Glossae and paraglossae are unequal in length. In specimens in alcohol the gill remnants are easily seen, but are more difficult to see in dried specimens.

8b Branched gill remnants absent from venter of thorax; simple and fingerlike gill remnants may arise from the submentum (Fig. 10A). **9**

9a Anal lobe of hind wing large and fanlike (Fig. 164); pronotum quadrate in shape; without a forked vein arising from basal anal (BA) cell of front wing. .... **Perlodidae**

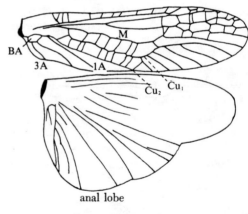

Figure 164

Figure 164   Perlodidae, wings

Nearly 100 species divided into about two dozen genera make up this common and widespread family. These are usually medium sized (5-15 mm). Adults are often greenish or yellow in life; some species are dark. Adults of some are known to feed on pollen. Submental gills are usually present (Fig. 10A).

9b Anal lobe (AL) of hind wing reduced or absent; front wing often with a forked vein arising from basal anal (BA) cell (Fig. 165); pronotum oval in shape. .... **Chloroperlidae**

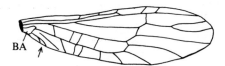

Figure 165

Figure 165 Chloroperlidae, wings; AL, anal lobe; BA, basal anal cell

There are 10 genera and over 50 species in this common and widespread family. Adults are often green or yellow, sometimes with a dorsal stripe, and are 5-15 mm in length. Glossae are much shorter than the paraglossae. Gills are never present.

AQUATIC HEMIPTERA
True Bugs

There are over 20,000 species of Hemiptera worldwide and about 4600 species in North America. About 10% of the species are aquatic. Of the total of 44 North American families, about 15 are aquatic or semiaquatic.

Aquatic Hemiptera are divided into two groups; the Cryptocerata have antennae which are short and hidden below the eyes (nearly all are swimming bugs); the Gymnocerata (shore bugs) have long and conspicuous antennae. In contrast to most other aquatic insects, adult and immature Hemiptera share the same habitat, habits, and general appearance. A separate key for adults and nymphs is not necessary. Body shape among aquatic Hemiptera shows extreme diversity, but the basic plan is similar. All have beak-like mouthparts, and leathery front wings which overlap at the tips; scent glands are usually present. Almost all Hemiptera overwinter in the adult stage, even in northern areas; nearly all are predatory, and all North American species are dependent on atmospheric air for respiration. (The Old World Aphelocleirus are able to live under water indefinitely.) Wing polymorphism (short, half-sized, and normal large wings) is common in aquatic Hemiptera, apparently being under both genetic and environmental control. While aquatic Hemiptera are highly diverse in appearance, they are considered to consist of a relatively closely related group of families, especially when compared to Orders such as Diptera and Coleoptera.

Both stages of Hemiptera can be preserved in alcohol. Adults may be mounted on pins.

1a Antennae inserted beneath the eyes and usually not visible from dorsal view (Figs. 17, 166) (but see Couplet 5b). .. 2

1b Antennae long, easily seen from above (Figs. 17, 175). 9

2a Beak fused to the head, not a distinct segmented tube; tarsi of front legs scoop-like (Fig. 166) (Water boatmen). .... **Corixidae**

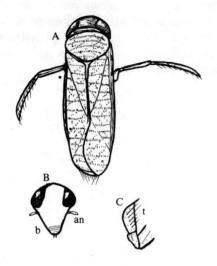

Figure 166

Figure 166 A. Corixidae; B. face view with antenna (an) and beak (b); C. front tarsus (t)

Water boatmen range from 3-10 mm in length and are among the most common aquatic Hemiptera. They are collected both in the water and at lights. The head is short and broad, the eyes are large, ocelli are absent, and the antennae are short and inconspicuous. The hind legs are oar-like. About 130 species and 20 genera are recognized in North America. Many species feed on particles of plant tissue but some are predatory on other small insects. Probably all species ingest some animal material. Many can produce sound by stridulating—rubbing the legs at the base of the front femur on the lateral margins of the head. Corixids break the water surface with the head and prothorax, picking up an air supply on the surface of the body and under the elytra. They can remain submerged for long periods of time. Corixids overwinter in the adult stage, often under ice.

2b Beak a definite 3 or 4 segmented tube or cone; front tarsi not scoop-like (Fig. 168). 3

3a Abdomen with two long tubes at the end (Fig. 167) (Water Scorpions). .... **Nepidae**

Figure 167

Figure 167 Nepidae (*Ranatra*)

There are but 3 genera with about a dozen species of water scorpions in North America. Ocelli are absent, the beak and antennae are 3 segmented, and the scutellum is conspicuous. The front legs are raptorial. *Ranatra*, the common genus, is long and thin as illustrated. It is 5 cm or longer when mature. *Nepa* is an eastern genus with a single species and the body is about 1/3 as wide as long. *Curicta* is subtropical. Nepids are deliberate and graceful insects in their medium but are clumsy and ungainly if they try to move rapidly or if they are taken out of water. They inhabit marshy and stagnant ponds where they lie in wait for any prey of appropriate size. Respiration is through the long respiratory tube. They are able to make sounds (stridulate) by jerking the legs and rubbing the front coxa along the margin of the coxal cavity.

3b Without long tubes at the end of the abdomen. 4

4a Toad-shaped shore bugs; ocelli present and conspicuous; middle and hind legs without a fringe of swimming hairs (Fig. 168). 5

4b Swimming or crawling water bugs, body shape and ocelli variable; middle and hind legs *usually* with a fringe of swimming hairs. 6

5a Eyes strongly protuberant; antennae hidden in grooves below eyes; front legs modified for grasping, differing from legs of middle pair (Figs. 17F, 168) (Toad Bugs). .... Gelastocoridae

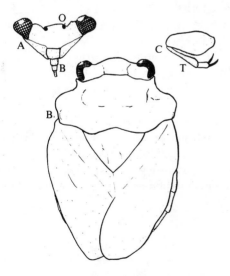

Figure 168

Figure 168 Gelastocoridae; A. Face view (ocellus, o); B. body without legs; C. front leg

There are two genera and under 10 species of toad bugs in North America. Ocelli are present, the antennae are hidden, and the beak is 4 segmented. They are often mottled yellow or brown and granulate in appearance. In *Gelastocoris* the tarsal segment bears a claw while *Nerthra* has the claw of the front leg attached to the tibia, the tarsus appearing to be absent. They live on mud or sand flats on the shores of streams or ponds where they run and leap rapidly and capture a variety of insect prey. Eggs are laid in the sand. They are common and widespread.

5b Eyes not as strongly protuberant and antennae easily visible in dorsal view; front legs not modified, similar to middle legs. .... Ochteridae

These are small bugs, 4-5 mm in length, which are blue or black in color. Both the beak and the antennae are 4 segmented. The front and middle tarsi are 2 segmented, the hind tarsus, 3 segmented. The legs are long and suited for running. They are rare but are distributed across North America along the shores of streams. They are predatory and seven species of the single genus *Ochterus* occur in North America.

6a Front legs flat and forming a pincer; dorsal side of insect flat (Figs. 169, 170). 7

6b Front legs round in cross section, do not form a pincer; dorsum convex (Figs. 171, 172). 8

7a Large insects, 2-5 cm in length; membrane of wing (arrow, Fig. 169) with a network of veins (Giant Water Bugs, Toe Biters) (See also Fig. 17H). .... **Belostomatidae**

Figure 169

Figure 169 Belostomatidae

These are some of the largest aquatic insects, sometimes exceeding 5 cm in length and 2 cm in breadth. Some species are smaller, about 2 cm or less. The ocelli are absent, the beak is stout and 3 segmented, and the antenna is 4 segmented but hidden under the eye. The front legs are very stout and pincer-like. The tip of the abdomen has a pair of retractile strap-like appendages used in respiration. They are attracted to lights and fly well. Most are tropical but there are 3 genera and about 20 species in North America. They overwinter as adults. In the genera *Belostoma* and *Abedus* eggs are glued to the backs of the males where they are carried until they hatch (Fig. 2F). They feed on water animals ranging from insects to tadpoles and small fish, and they can inflict a painful "bite" with the beak on the careless collector.

7b   Smaller, about 1 cm or less in length; membrane of wing without veins (arrow, Fig. 170) (Creeping Water Bugs). .... **Naucoridae**

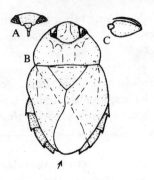

Figure 170

Figure 170 Naucoridae; A. Face view of head; B. body without legs; C. front leg

While distantly related to the giant water bugs, Naucorids superficially resemble them. Naucorids lack the strap-like appendages at the tip of the abdomen of the Belostomatids. The smooth oval outline of the Naucorid body is characteristic. There are 4 genera and about a dozen species in North America. They can inflict a painful "bite" with the beak. Naucorids live in quiet water where they creep through submerged vegetation searching for prey. They are often found in hot or warm springs.

8a   Body oval and less than 3 mm in length; wing membrane absent; two apical claws on hind leg (Figs. 17E, 171) (Small Backswimmers). .... **Pleidae**

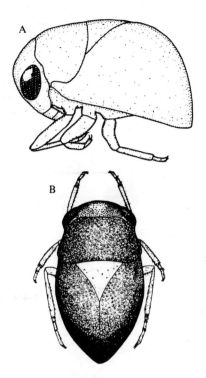

Figure 171

Figure 171   Pleidae; A. lateral view; B. dorsal view

These backswimmers are tiny (2 mm), seed-like insects with humped backs. The eyes are small, the beak is 3 segmented, and the tarsi are 2 segmented and lack swimming hairs. Most members of the family are tropical, but the genus *Plea* with seven species is widespread in North America. They are found in masses of vegetation, usually crawling rather than swimming. They feed on ostracods and other small crustacea.

8b   More than 5 mm in length; form more slender and elongate; wing membrane present; hind leg with one apical claw (Fig. 172) (Backswimmers). .... **Notonectidae**

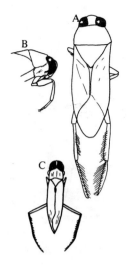

Figure 172

Figure 172   A. Notonectidae (*Notonecta*); B. Side view of head and front legs; C. *Buenoa*

These are elongate, deep bodied, common insects with large eyes, no ocelli, concealed antennae, and a 4 segmented beak. The hind legs are elongate with hairs for swimming and the front legs are adapted for holding prey. Ventrally there is a keel of hairs that traps air for respiration when the insect breaks the surface. There are two genera and about 50 species in North America. Along with the Corixidae these are the most commonly seen aquatic Hemiptera. *Buenoa* are 5-8 mm long and have an elongate pit dorsally just behind the scutellum (Fig. 172C); this is lacking in the 10-15 mm long *Notonecta* (Fig. 172A). These are predators which swim on their backs searching for prey in relatively open water.

9a   Body greatly elongated, head more than 1/4 the total body length (Fig. 173) (Marsh Treaders). .... **Hydrometridae**

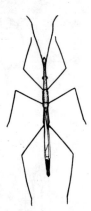

Figure 173

Figure 173   Hydrometridae

These are scarcely 1 cm long insects with slender bodies and hair-thin legs. They are very
difficult to see while collecting but are not
rare. The antennae are 4 segmented and the
tarsi are 3 segmented with terminal claws.
Hydrometra, the only genus in North America,
consists of about 10 species, with widespread
distribution. These peculiar sluggish insects
occur on the surface of quiet water protected
by emergent vegetation. They are said to feed
on small crustaceans and mosquito larvae
which they spear with the beak through the
surface film. They overwinter as adults.

9b **Body not greatly elongated, body shape
not as in Fig. 173.** **10**

10a **Claws arise from the tip of the tarsus
(Fig. 180A, arrow).** **12**

10b **Claws arise from the bottom of the tarsus and not from the tip (Fig. 175A, ar-
row).** **11**

11a **Hind legs very long, femur extends beyond the tip of the abdomen (Fig. 174,
arrow) (Water Striders).** .... **Gerridae**

Figure 174

Figure 174   Gerridae (femur, arrow)

Gerrids are the long-legged water striders
known to almost everyone. The beak and antennae are 4 segmented and a scutellum is absent. The hemelytra lacks a membrane. In
North America there are 12 genera and about
60 species. *Gerris* is the most common genus
and adults are 1-2.5 cm in length. Species in
other genera may be but 2 or 3 mm in length.
Halobates lives on the ocean surface, often
hundreds of miles from land. All gerrids are
surface inhabiting predators that feed on insects and anything else they can catch. The
front legs are modified for grasping prey, the
middle pair are used for locomotion, and the
hind pair are used for steering.

11b **Hind legs not elongated, femur not extending beyond the tip of the abdomen
(Fig. 175) (Riffle Bugs, Broad Shoul-
dered Water Striders).** .... **Veliidae**

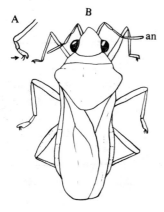

Figure 175

Figure 175 Veliidae, A. Detail of subapical claw (arrow); B. body shape

These are small insects often ranging from 1 to 5 mm in length. The beak and antennae are 4 segmented, the scutellum is absent. Their close relationship to the Gerridae is indicated by the subapical claws of both. The common names are descriptive of both their habitat and appearance. There are 5 North American genera and about 50 species. *Microvelia* are common pond forms while *Rhagovelia* are associated with streams. The latter have a remarkable plume of hairs on the middle tarsi. Veliids are surface inhabiting predators which can capture mosquito larvae and other small animals through the surface film. They overwinter as adults. The group is widespread and common but not seen as frequently as Gerridae. See also Fig. 17D.

12a Femur of the hind leg attached to the abdomen by one short segment (trochanter, the basal segment (coxa), forming a flat, triangular plate fused to the abdomen (Fig. 176). 13

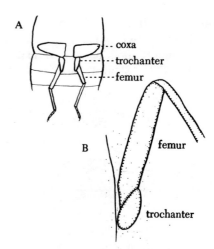

Figure 176

12b Femur of hind leg attached to the abdomen by two short segments (coxa and trochanter). 14

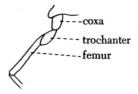

Figure 177

13a Eyes bulging; ocelli closer to each other than to the compound eyes; all antennal segments about equal in thickness (Figs. 17I, 178) (Shore Bugs). .... Saldidae

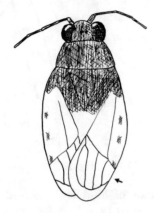

Figure 178

Figure 178 Saldidae

These are 5-8 mm long and often marked with black and white. Saldids are active bugs found on sand or mud shores of various types of aquatic habitats. The antennae are 4 segmented and the beak 3 segmented, and the eyes are notched on the inner margin. The wing membrane (arrow) usually has 4 or 5 closed cells, quickly separating this family from others similar in appearance. The family in North America consists of about 10 genera and 60 species. They are active predators and require effort to collect because of their agility. They are common.

13b Eyes not notably bulging; ocelli very close to the compound eyes, 3rd and 4th antennal segments thinner than segments 1 and 2 (Jumping Ground Bugs). .... Dipsocoridae

Some species in this family are semiaquatic, others are terrestrial. They are rarely collected but apparently widely distributed (records are from Georgia and California). They live along steams where they move with ease through the crevices in wet gravel.

14a Body length about four times as long as body width; tarsi with 3 segments. 15

14b Body length about twice as long as body width; tarsi with two segments (Figs. 17G, 179) (Velvet Water Bugs). .... Hebridae

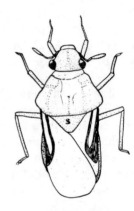

Figure 179

Figure 179 Hebridae; s—scutellum

These are tiny, squarish, velvety insects not exceeding 2.5 mm in length with eyes that are coarsely faceted; the antennae are 4 or 5 segmented, the scutellum (s) is present and the pronotum appears collarlike. The hemelytra is mostly membranous and the membrane is without veins. When folded the beak extends nearly to the base of the hind legs. There are two North American genera, *Hebrus* with 5 antennal segments and *Merragata* with 4. They are widespread and common but seldom seen because of their small size. They are found on mats of green algae, on floating plants, in debris piles cast up by waves, or near the shores of stagnant ponds. In collecting, this material can be submerged and the bugs will float to the surface.

15a Tibia of hind legs with stiff black bristles, the length of which nearly equals the diameter of the tibia (Fig. 180) (Water Treaders). .... Mesoveliidae

15b Tibia of hind legs with short hairs, the length of which equals 1/3 or less the diameter of the tibia. ...... Macroveliidae

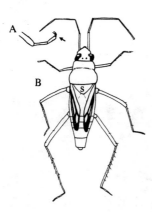

Figure 180

Figure 180   Mesoveliidae, A. detail of apical claws; B. body shape, s—scutellum

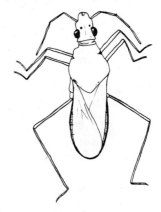

Figure 181

Figure 181   Macroveliidae

There is the single widespread and common genus, *Mesovelia,* with 4 species in North America. They are 2 to 5 mm in length and are greenish. The wings when present are black-veined and membranous; the antennae are 4 segmented and the beak 3 segmented. The scutellum is visible (s). They inhabit the water surface or live on surface vegetation. Overwintering is in the adult stage. Even though the body is green, suggesting plant feeding, they are said to be predatory on insects of the surface and also on ostracods which they spear through the surface.

This family contains the single genus and species *Macrovelia horni* which is widely distributed in the west (Arizona, New Mexico, California to the Dakotas and Colorado) but is rarely collected. It has at times been included in the Mesoveliidae or Veliidae, but its unique family rank is now generally accepted. *Macrovelia* has veins and cells in the membrane of the hemelytra and the pronotum projects posteriorly to cover the scutellum. They have well developed ocelli and apical claws. They are usually found on moss, debris, or vegetation along shores of springs and streams and do not occur in or on open water. Their biology is poorly known.

TRICHOPTERA
Caddisflies

Trichoptera are a medium-sized group with about 4500 species worldwide and about 1000 species and 17 families in North America. Adults (Figs. 39, 199A) are brownish mothlike insects which range from a few mm to nearly

3 cm in length. Larvae are of two types, sometimes referred to as cruciform (Fig. 183), which are case builders with anal legs fused to the abdomen, and campodeiform (Fig. 191), which are free living larvae with elongate anal

legs. The most spectacular features of the Trichoptera are the cases and food capturing nets constructed by the larvae (Figs. 31, 182-184, 187). Larvae can be found in all habitats except that they are not tolerant of excessive pollution. Adults can be found near the larval habitat, usually crawling over vegetation. All stages are best preserved in alcohol.

TRICHOPTERA LARVAE

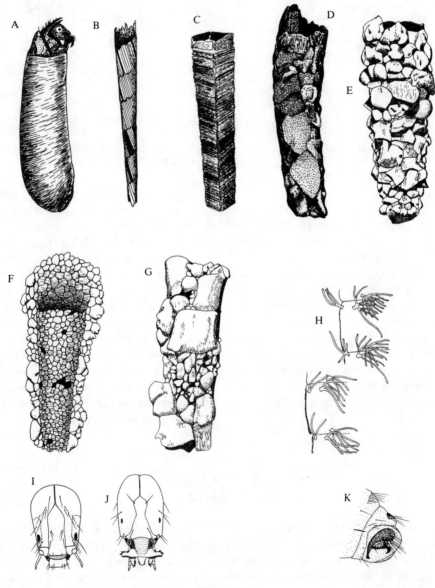

Figure 182

Figure 182   Caddisfly cases (A-G), head capsules (I, J) and abdominal gills (H) and anal legs (K)

1a   Larva with a case in the shape of a snail shell (Fig. 183C); claw of anal legs as in Fig. 183B, not hook-shaped, but comb-like. .... **Helicopsychidae**

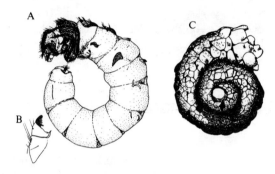

Figure 183

Figure 183   Helicopsychidae, A. larva; B. anal claw; C. case

There is a single genus, *Helicopsyche*, with 4 species in North America. They are widespread and common, occurring in streams and rivers, on wave-washed lake shores, and in warm springs. The case (6 mm or less in diameter) is made of close fitting rock fragments, giving it a smooth and neat appearance. They were first described as a species of snail (in 1834). There is an opening in the spire of the case which allows water circulation for respiration. Larvae often live under stones or in the spaces in loose gravel substrate and may burrow up to a foot below the surface. Larvae feed on plant material. Large aggregations of pupae and their spiral cases are sometimes found on rocks.

1b   Larval case not in the shape of a snail's shell; anal claw hook-shaped (but may have teeth). 2

2a   Small, gill-less, case-building larvae, under 5 mm in length, with a dorsal sclerotized shield on each of the thoracic segments (Fig. 184). ....
.... **Fifth instar Hydroptilidae**

Note: Two problems may arise here. First instar Hydropsychidae will key to this Couplet, but can be separated by the brush of setae on the anal legs (Fig. 185, arrow).

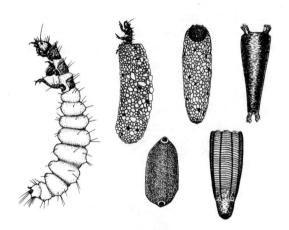

Figure 184

Figure 184   Hydroptilidae, larvae and various cases

Also, instars 1-4 of Hydroptilidae do not construct a case. Their appearance is as in Fig. 184 except that the membranous parts of the body are not expanded, the abdomen not exceeding the thorax in length or diameter. As is seen in Fig. 184, the abdomen of a 5th instar larvae is greatly expanded. These changes in appearance, unique to this family, are considered to be hypermetamorphosis.

These larvae occur in lakes, streams, and rivers where they feed on algae and diatoms. The numerous species are widespread and common but usually overlooked by collectors because of their small size. This is a very diverse group. Some construct cases which are "bivalve" in shape, others are bottle shaped, and others are typical tube-like cases of fine sand (Fig. 184). Cases are often attached to rocks. Early stage larvae should be recognizable by the body size of 1 or 2 mm, the three thoracic shields, the short abdomen which is sclerotized dorsally, and by the lack of setal brush on the anal legs. Early instar larvae are poorly known and seldom collected, but apparenty occupy the same habitat and consume similar food to the 5th instar.

2b   Larva not as described; if three dorsal thoracic shields are present, then ventral gills are also present; size variable; case present or absent. 3

3a   Larva with three dorsal thoracic plates *and* with ventral gills; larva without a movable case; body up to 2 or 3 cm length (Fig. 185). .... Hydropsychidae

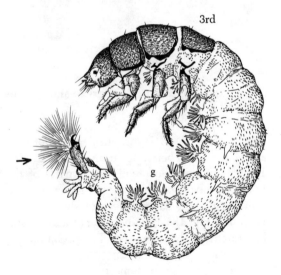

3rd

g

Figure 185

Figure 185   Hydropsychidae with 3rd thoracic plate; arrow, brush on anal leg; g, gills

These larvae live in silk and stone retreats under rocks or in debris in flowing water, or oc-

casionally on wave-washed shores. There are 11 genera and about 145 species in North America and they are very common and widespread. In addition to the key characters, larvae have a conspicuous brush of hairs on each anal leg (arrow, Fig. 185). Larvae secrete silken nets which filter algae, organic particles, and small organisms from the passing current (Fig. 187B). Just prior to pupation a case-like cell is constructed. Larvae are able to produce sound by rasping the femur of the front leg over ridges on the underside of the head. These larvae may be extremely abundant in favorable conditions.

3b   Larva with but one or two dorsal plates behind head; gills and case variable. 4

4a   Larva with a portable "tortoise-shell" case, case is a dome-like mound of stones with a ventral strap and open at both ends; 8 mm or less in length. .... Glossosomatidae

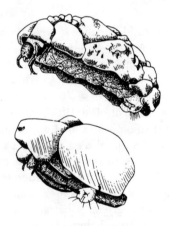

Figure 186

Figure 186   Glossosomatidae, larvae in cases

In the absence of a case, larvae can be recognized by the single complete dorsal plate behind the head, a dorsal sclerotized plate on abdominal segment 9, anal legs broadly fused to the abdomen, anal claws with at least one dorsal accessory hook, and the absence of a prosternal horn. Larvae live in rivers and streams and are widespread and abundant. They graze on algae and diatoms on the upper surfaces of stones. The stones of the case have openings between them to allow circulation of water. The larvae lack gills. While there are about 80 species, both larvae and their cases are very uniform in appearance.

4b   **Case, if present, not "tortoise shell" in structure.** **5**

5a   **Larva without a portable case; end of abdomen with a distinct pair of elongate anal legs with claws (Figs. 185, 187A, 191A).** **6**

Larvae of this group are either free-living predators, or they construct silk nets in retreats under rocks or debris (Fig. 187B). They lack a prosternal horn, have a single complete sclerotized plate behind the head, and the anal claws lack dorsal accessory hooks.

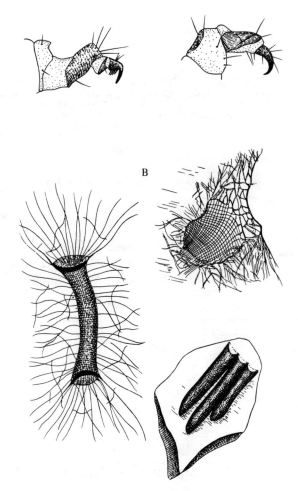

B

Figure 187

Figure 187 A. Anal legs and claws; B. various nets of Trichoptera

5b   **Larva with a case; anal claws appear to be attached directly to the abdomen (Figs. 188, 197, arrows).** **8**

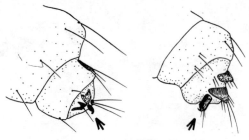

Figure 188

Figure 188 End of abdomen, and legs and claws (arrow)

6a   With a distinct sclerotized dorsal plate (p) on the segment that bears the anal legs (Fig. 189, p). .... Rhyacophilidae

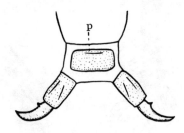

Figure 189

Figure 189 Rhyacophilidae, anal legs with plate (p) on segment 9

Body shape resembles Fig. 191. In some specimens it may be necessary to examine segment 9 very carefully because the plate may not be pigmented and conspicuous. The area will appear shiny. Larvae have a single dorsal plate behind the head. The thoracic legs are all of similar size. Larvae may have gills, which may be single or bushy. *Himalopsyche* with a single species and *Rhyacophila* with over 100 species, occur in North America. Larvae are free living, having neither a portable case nor a fixed retreat. A cell is constructed prior to pupation. They are mostly predators but in many species plant material is found in the gut contents. Mature larvae of *Himalopsyche phryganea* exceed

30 mm in length and adults also reflect this size; *Rhyacophila* are usually 15-20 mm long when mature. Both genera may have bushy gills on the abdomen, but only *Himalopsyche* has them on the thorax.

6b   Without a dorsal sclerotized plate on the segment that bears the anal legs (segment 9). 7

7a   Labrum membranous, anterior part expanded laterally to form a T shape (Figs. 182J, 190). .... Philopotamidae

Figure 190

Figure 190 Philopotamidae, head and labrum (1)

These larvae lack a case but construct nets of silk in streams and rivers. The body is slim (resembling Fig. 191A), elongate, white, and lacks gills on the abdomen. There is a dorsal plate only on the prothorax. This plate is often edged with a distinct black line. Both the plate and the head are light brown. Larvae spin silk sac-like nets in spaces under rocks and fine particles are filtered from the current. This food material is cleaned from the net with the peculiar T-shaped labrum. There are three widespread North American genera and about 40 species.

7b   Labrum sclerotized and rectangular in shape (Fig. 191B). 18

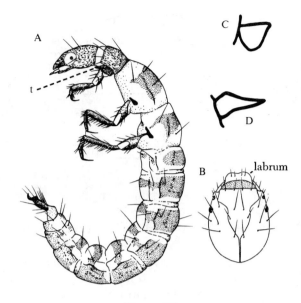

Figure 191

Figure 191 Polycentropodidae, A. Body shape; B. head; C. trochantin of Psychomyiidae and D. Polycentropodidae. t—trochantin (see Couplet 18)

8a Antenna relatively long, last segment 3 to 6 times as long as wide (Figs. 182I, 192A); usually with an unpigmented line below the eye (Fig. 192A, arrow). .... **Leptoceridae, in part**

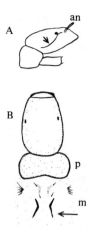

Figure 192

Figure 192 Leptoceridae, A. Head with antenna (an) and subocular line (arrow); B. head and thorax with comma-shaped sclerites (arrow). Prothorax (p) and mesothorax (m) of *Ceraclea*

This is a large and diverse group with over 100 species in seven genera in North America. All but one genus is characterized by long antennae (see Couplet 9a). All have a more or less distinct unpigmented line below the eye. The hind legs extend forward over the head and are longer and thinner than the front and middle legs. The hind legs in some genera have long fringes of hair and are used for swimming. Leptocerids inhabit both flowing and still water and are very common and widespread. Cases vary and include flattened tubes with lateral flanges, cases of pure silk which are often tapered and curved, coarse textured tubes of rock mixed with plant material, and neat tapered cases of spirally arranged plant fragments (compare Fig. 182A, B, D, F). One group is associated with freshwater sponges (see Couplet 9b), others are notable because they can tolerate very high temperatures, and some are abundant enough to be pests of cultivated rice. Food is primarily plant material, but some feed on small aquatic animals.

8b Antennae very short and difficult to see, about twice as long as wide and appearing as little more than a small bump between the eye and base of the mandible (Figs. 197, 198). 9

9a Mesonotum very lightly sclerotized and lightly pigmented, and with a pair of dark curved markings on each side on the posterior half (Fig. 192B, arrow). .... **Leptoceridae, in part**

See also Couplet 8a. These are Leptoceridae of the genus *Cercalea,* some species of which have short antennae, possibly as an adaptation to living and feeding on freshwater sponges. (Not all members of the genus live on sponges). The subocular line is usually visible (Fig. 192A). The case is of nearly pure silk in the sponge-inhabiting species but includes sand and often has a lateral flange or overhanging lip in non-sponge feeding species. They are common and widespread.

9b Mesonotum not as described, either entirely membranous or with sclerites in various shapes. 10

10a Labrum with a distinct row of stout setae (Fig. 193); case either a hollowed twig or piece of wood, or of leaf fragments. .... **Calamoceratidae**

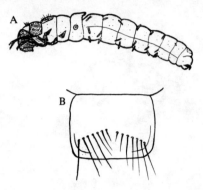

Figure 193

Figure 193 A. Calamoceratidae; B. front view of labrum with setae

This group is largely tropical, but 3 genera and 5 species are found north of Mexico. Abdominal gills, a lateral fringe, and dorsal and lateral humps of the first abdominal segment are present. The anterolateral angles of the pronotum form a point. *Anisocentropus* (1 species) occurs in the southeast U.S. and constructs a case with a cut leaf for a dorsal shield; *Heteroplectron* has an eastern and a western species and cases are of excavated twigs or wood; *Phylloicus* with 2 species occurs in Texas and Arizona and the case is primarily of leaf fragments or wood and bark. All larvae live in running water but may be confined to pool areas. They feed on plant material and development appears to require more than 1 year.

10b Larvae with no more than six setae on the labrum and setae not in a row; case variable. 11

11a With a single sclerotized plate behind the head; i.e., meso- and metanotum entirely membranous (Fig. 194) (See Note). .... **Phryganeidae**

(*Note:* The genus *Yphria,* with a single species found in Oregon and California, has the mesonotum about half covered by a sclerite.

The case is up to 3 cm long and has a ventral strip of rock fragments. Other characters are as follows.)

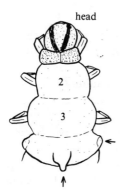

head

Figure 194

Figure 194 Phryganeidae, membranous meso- and metanotum (2, 3) and tubercles (arrows) of the first abdominal segment

But for the above exception, this is the only group which has the combination of a prosternal horn and a single scleritized plate behind the head. In addition, there is a dorsal sclerotized plate on segment 9. Gills, a well developed lateral fringe, and spacing tubercles (abdomen segment 1) are present; the head capsule and sometimes the membranous areas of the thorax show striking color patterns. Larvae may be large, sometimes over 30 mm in length. Cases are made of leaves and other plant material and are either spiral in construction (compare Fig. 182B) or of leaf fragments carefully fitted together end to end so that the case consists of a series of rings. Others may have irregular cases. Habitats vary from marshes, lakes, and temporary pools to flowing water. They are most abundant in the northern latitudes. In North America there are 10 genera and 27 species.

11b With small or large sclerotized plates on the meso- and metanotum. 12

12a Claws of hind legs different in structure from those of middle and front legs, either short and stub-like with bristles (*Molanna*) or elongate and with setae (*Molannodes*). .... Molannidae

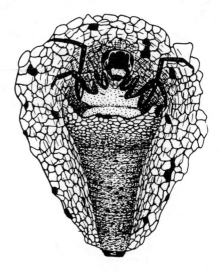

Figure 195

Figure 195 Molannidae, ventral view of case and larva

This family can usually be recognized by the peculiar cases which are of fine sand grains with lateral flanges (Fig. 195); the case is rather fragile and its flanges may be lost. This is a small group of two genera and seven species which are eastern and northern in distribution. *Molannodes*, with a single species and long hind tarsal claws and often with plant material in the case, is restricted to Alaska and the Yukon as well as Asia and Europe. *Molanna* has stubby rear claws and is widely distributed in the east. The genus extends across Canada and south in the mountains to Colorado. Both genera live on sand or mud in lakes or slow areas of rivers, and some are found in cold springs and in deep water in lakes. They feed primarily on plant material. They are common in favorable habitats.

12b Claws of hind legs not as described, similar in structure to claws of front and middle legs; if hind claws are elongate, see Couplet 14a. 13

13a First abdominal segment without dorsal and lateral humps (contrast with Fig. 194, arrows); pronotum with a transverse ridge and groove (Fig. 196, g); case usually either 4-sided in cross section or a smooth round tube of silk or sand grains (Figs. 196, 182C). ....
.... **Brachycentridae**

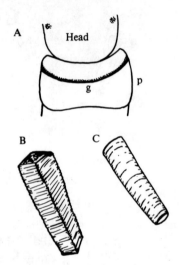

Figure 196

Figure 196 Brachycentridae, A. Pronotum (p) with groove (g); B, C, cases

There are six genera and about 30 species in North America. They are widespread and common, inhabiting flowing waters ranging from cold springs to marshy rivers. In this group the abdominal gills and lateral fringe are reduced or lacking and the prosternal horn may be present or absent. Most feed on plant material, but some are known to be predaceous.

13b Dorsal or lateral humps usually conspicuous and pronotum without a deep transverse groove; case variable. 14

14a Rare and eastern, case 6 mm or less in length and larvae with hind claws long and tapered, pronotum with a distinct ridge leading to an antero-lateral projection. .... **Beraeidae**

In this group the case is curved and tapered and made of smooth fine sand grains. Larvae have dorsal and lateral humps on abdominal segment 1, there are two stout apical setae at the end of the abdomen, and there are two clusters of about 30 fine setae between the anal legs. There is a single genus, *Beraea*, with 3 species which are rare but sometimes locally abundant in the east. They live in small springs or streams or in the mud of seepage areas, where they feed on plant material. Two years are probably required to complete larvae growth.

14b Case and larva not as described. 15

15a Antenna very close to the eye (contrast with Fig. 198); dorsal humps on abdominal segment one lacking (Fig. 197). .... **Lepidostomatidae**

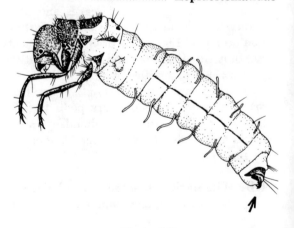

Figure 197

Figure 197 Lepidostomatidae; arrow—anal leg and claw

There are two genera and about 70 species of this widespread and common group in North America. They resemble Limnephilidae except for characters in the key. Lepidostomatids also lack ringed chloride epithelia on the abdominal segments (see Couplet 16a), the prosternal horn is present, abdominal gills are single and arranged in dorsal and ventral rows, or are absent, and the lateral fringe is sparse or absent. Larvae sometimes construct square-sided cases or they may be irregular. Cases are usually of plant material. They live in flowing water or on lake shores and feed on plant debris.

15b Antenna midway between the eye and base of the mandible, *or* very close to the mandible (Fig. 198); dorsal hump present on abdominal segment one. 16

16a Common; antenna located about half way between the eye and the base of the mandible (Fig. 198); chloride epithelium (see below) usually present on ventral abdomen; prosternal horn present, case variable. .... **Limnephilidae**

(Chloride epithelia are areas of modified cuticle, usually oval areas on the ventral abdomen, bordered by a thin sclerotized line. They are organs of osmoregulation and ion exchange. They are found in Limnephilidae, Hydroptilidae, and Molannidae.)

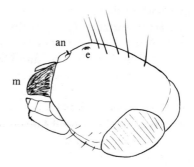

Figure 198

Figure 198 Limnephilidae, side view of head; m, mandible; an, antennae; e, eye

Limnephilidae occupy a wide range of habitats. The cases are extremely variable. They may be of either plant or mineral material, and are sometimes smooth, tapered, neat or very irregular in construction (Figs. 31, 182). There are over 50 genera and 300 species that are distributed throughout North America, especially the northern latitudes. Larvae often have numerous branched or single gills on the abdomen (Fig. 182H). The prosternal horn is present but may be small; the lateral fringe is usually well developed. With over 50 genera in the group there are exceptions to almost every character except the location of the antenna as indicated in the key. Larvae feed on plant material. There is usually one generation per year.

16b Uncommon; antenna very close to the base of the mandible; prosternal horn never present; case always of sand or mineral material. 17

17a Anal proleg with a cluster of about 30 setae located above and inward from it; front trochantin hook-shaped (compare Fig. 191). .... **Sericostomatidae**

A recent revision places 3 genera (*Gumaga, Fattigia,* and *Agarodes*) and 12 species in this group. Larvae are found in flowing water or on wave-washed shores where they often burrow under the surface. They are found from Florida to Maine in the east and from New Mexico and Utah to California and Oregon in the west. Larvae feed on fine organic particles.

17b **With only about 5 setae near each anal leg; front trochantin not hook-shaped.** .... **Odontoceridae**

Based on recent studies this family contains 12 North American species and 6 genera (*Pseudogoera, Parthina, Psilotreta, Marilia, Nerophilus,* and *Namamyia*). Larvae of this group burrow in the substrate of flowing water and they are seldom collected unless this habitat is searched. They are scavengers which feed primarily on plant material. The cases are very resistant to crushing. *Psilotreta*, with 7 eastern species, is distributed from Tennessee to Quebec and Ontario and west to Wisconsin. Others are known from the west, including Oregon, California, and Texas.

18a **Trochantin of front leg acute and sharp at the apex (Fig. 191A (t), D).** .... **Polycentropodidae**

These larvae are in part predatory. They construct nets which may be tube-like or trumpet-shaped, and fine particles are filtered from the water as food. Some occur regularly in lakes where body undulations by the larva causes a current to flow through the net.

18b **Trochantin of front leg broad, hatchet-shaped (Fig. 191C).** ...... **Psychomyiidae**

These larvae construct debris or sand-covered silk tubes on the surfaces of rocks or on wood. They rarely occur in still water. Food is mainly detritus and algae.

TRICHOPTERA ADULTS

Figure 199 A. Trichoptera adult in resting position; B—N. Various structures, see keys. m—mesoscutum, s—mesoscutellum, max-maxillary palp, lab—labial palp, a—apical spur, p—preapical spur

1a **Body small, often 2 or 3 mm in length; antenna shorter than the wings or body; mesoscutum (m) without warts and with posterior portion (mesoscutellum, s) forming a more or less triangular area with steep sides posteriorly; wings usually with fringe hairs longer than the breadth of the wing (Fig. 200).** .... **Hydroptilidae**

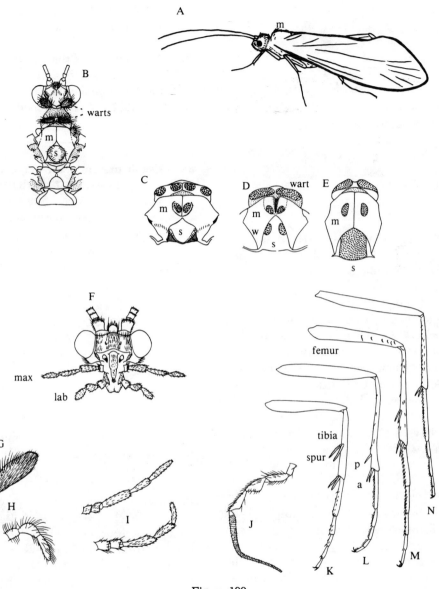

Figure 199

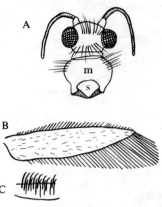

Figure 200

Figure 200 Hydroptilidae; A. Head and thorax; B. front wing; C. detail of hairs on wing. m—mesoscutum, s—mesoscutellum

Body length ranges from 2-6 mm, and while Hydroptilidae are often called microcaddisflies, members of other families such as Glossosomatidae and Psychomyiidae may fall in this size range. Fourteen genera and about 175 species occur north of Mexico. They are widespread and common but often overlooked in both the adult and larval stages. In this group the maxillary palpi are 5 segmented, the body is very hairy, the front tibia never has more than 1 spur, the mesoscutellum lacks warts, the front wings have semi-erect club-like hairs (Fig. 200C). The wing itself is pointed (Fig. 200B). Ocelli may be present or absent. They are most common near flowing water.

1b   Size usually larger and not with the above combination of characters; mesoscutellum evenly convex, not steep-sided posteriorly, and often with warts; antenna usually as long as body or longer; fringe hairs not longer than the breadth of the wing. 2

2a   Last (5th) segment of the maxillary palp either about twice as long as the 4th segment, or 5th segment flexible and differ-

ent in structure from the preceding segment (Figs. 201, 202, 199J). 3

2b   Last segment of the maxillary palp about the same length as the preceding segment and of similar structure (Figs. 203-208, 199F, I), or palp modified as in Figs. 199G, H. 5

3a   Ocelli present (but may be hidden by pubescence); last (5th) segment of maxillary palp elongate and flexible but annulations not conspicuous (Fig. 201), fourth segment shorter than the third. .... Philopotamidae

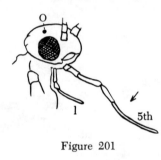

Figure 201

Figure 201 Head of Philopotamidae, o—ocellus; l—labial palp, arrow—maxillary palp with 5th segment

In addition to the above characters, the spur count is 2-4-4, 1-4-4, or 0-4-4, the tibiae and tarsi of the females may be flattened and dilated, and the end of the abdomen in females is prolonged into a conical ovipositor (also found in other families such as Hydroptilidae, Rhyacophilidae, and Glossosomatidae). While widespread and common, adults of this group are secretive and often must be searched for in vegetation overhanging streams and rivers. Three genera and about 40 species are known from North America.

3b   Ocelli absent from the top of the head, 5th segment of maxillary palp with annulations (Fig. 202, 199J). 4

4a Mesoscutum with warts as in Figs. 199E, 211 (arrow). **21**

4b Mesoscutum (m) without warts (Fig. 202C). .... **Hydropsychidae**

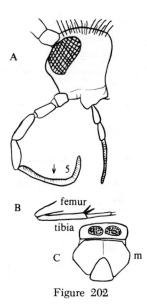

Figure 202

Figure 202 Hydropsychidae, A. head; B. front leg; C. thorax arrow—maxillary palp, 5— 5th segment, m—mesoscutum

In this group ocelli are absent, the maxillary palpi of both sexes are 5 segmented, the last segment being elongate and marked with cross striations, the preapical spur of the front leg is absent, the middle leg of the female is often flattened and dilated, and the mesoscutellum lacks warts. They are found near flowing water. In many areas this is the most abundant and common group of caddisflies. They range from 6 to 20 mm in body length. This family is characteristic of large rivers. Adults often occur in such numbers as to be a nuisance and have been the object of control measures, especially where the larval habitat has been enriched with sewage.

5a Ocelli (o) present (Figs. 201, 203). **6**

5b Ocelli absent (Figs. 202, 211). **11**

6a Maxillary palpi with 3 segments (Fig. 203). .... **Male Limnephilidae**

See Couplet 10b.

Figure 203

Figure 203 Male Limnephilidae; o—ocellus; 3—3rd segment of maxillary palp

6b Maxillary palpi with 4 or 5 segments. .. **7**

7a Maxillary palpi with 4 segments (Fig. 204). .... **Male Phryganeidae**

See Couplet 10a.

Figure 204

Figure 204 Male Phryganeidae; 4—4th segment of maxillary palp

7b   Maxillary palpi with 5 segments. 8

8a   First two segments of maxillary palpi short and about equal to each other in length (Figs. 205, 206). 9

8b   First two segments of palpi with second much longer than the first (Fig. 207). ....
.... **10**

9a   Size large, body length usually 12-24 mm; front tibia with a preapical spur (Fig. 205B). .... Rhyacophilidae

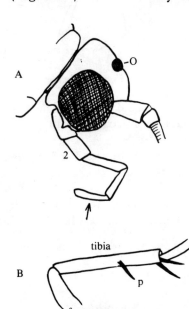

Figure 205

Figure 205 Rhyacophilidae, A. Head; B. front tibia. o—ocellus, 2—2nd segment of maxillary palp, p—preapical spur

Ocelli are present, the maxillary palpi are 5 segmented and similar in both sexes, and the

spur count is 3-4-4. A single species of the very large bodied *Himalopsyche* is known from California, Washington, and Oregon. About 100 species of the widespread and common genus *Rhyacophila* are found in North America.

9b   Body length not longer than 10 mm, sometimes as small as 3 mm; front tibia without a preapical spur (Fig. 206). ......
.... **Glossosomatidae**

Figure 206 Glossosomatidae, A. Head; B. front tibia; C. abdomen with plate-like structures

Six genera and about 80 species are known from the U.S. and Canada. Body length ranges from 3-10 mm. Glossosomatids often have spines or plates ventrally on segments 6 and 7 (as do rhyacophilids) and the female abdomen is elongate and tapered, often with two terminal cerci.

10a   Front tibia with two spurs, middle tibia with 4 spurs as in Figs. 199M, N. ....
.... **Female Phryganeidae**

See also Couplet 7a. There are 10 genera and 27 species of this family in North America. The family is widespread and common and usually easily recognized. Adults are often found near ponds or lakes as well as flowing water. As in the Limniphilidae, the legs often have a number of black spines as well as spurs and hairs; ocelli are present. Size varies from medium to large, to nearly 3 cm. The spur count is 2-4-4. Maxillary palpi are different in the two sexes.

10b   Front tibia with 1 spur, middle tibia with 2 or 3 spurs (Fig. 207B, C). ....
.... **Female Limnephilidae**

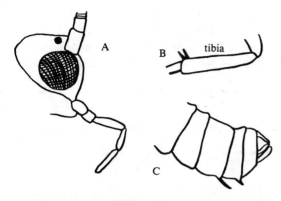

Figure 206

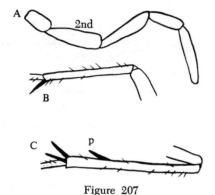

Figure 207

Figure 207 Female Limnephilidae. A. Maxillary palp; B. front leg with single spur on tibia; C. middle leg. p—preapical spur

See also Couplet 6a. Limnephilids are medium to large in size (1-3 cm body length); ocelli are present, the maxillary palpi of males are 3 segmented, of females, 5 segmented, with the fourth segment more than half as long as the fifth and the second much longer than the first. The tibiae and tarsi often have stout black spines and there is usually a black spot on the second leg segment (trochanter) which is easily seen from ventral view. This is the largest family of caddisflies in North America

and the group is most abundant in the northern latitudes. There are 52 genera and over 300 species. Adults may be found near water of all types, especially ponds, reflecting the wide range of habitats occupied by the larvae.

11a Antenna usually twice or more the length of the wings; mesoscutum (m) with a long line of seta-bearing warts (Fig. 208B, arrow). 12

11b Antennae about as long as the wings; warts not in a long series. 13

12a Scape of antenna 3 times longer than pedicel; tibia with a row of black spines (Fig. 208C); common. ...... Leptoceridae

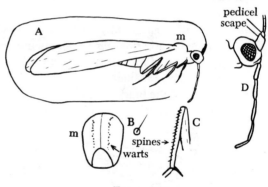

Figure 208

Figure 208 A. Leptoceridae; B. thorax; C. middle tibia; D. head, palps and antenna, m—mesoscutum

These are medium-sized caddisflies (10-15 mm) which lack ocelli and which have antennae 2 or 3 times the length of the body. The maxillary palpi are 5 segmented. The middle tibia lacks preapical spurs. This family is widespread and common, and north of

Mexico there are 7 genera and about 100 species.

12b Scape of antenna about as long as pedicel; tibia without a row of black spines (as in Fig. 207C); uncommon. .... Calamoceratidae

The middle tibia has a preapical spur. In North American representatives, the antennae are two to three times the length of the wings. Three genera and about half a dozen species are known from N.A. *Anisocentropus* (= *Ganonema*) is found in the southeast. *Phylloicus* (= *Notiomyia*) is found in the southwest. *Heteroplectron* is found from California to Oregon, B.C., and Alaska and also is known from New York and Quebec and southward.

13a Front tibia apparently with 1 spur; hind wing with anterior margin somewhat cut away beyond middle; with a row of stiff hairs or hooks(hamuli) along the basal portion of the hind wing (Fig. 209). .... Helicopsychidae

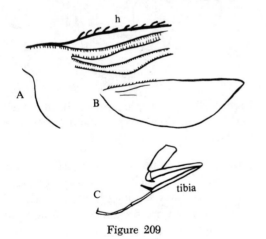

Figure 209

Figure 209 Helicopsychidae. A. Base of hind wing; B. hind wing; C. front leg. h—hamuli

There are 4 species in the genus *Helicopsyche* in our area, with the species *H. borealis* being

widespread and common, extending coast to coast and north into Canada. Body length is about 6 mm. Maxillary palpi of the males appear to be 2 segmented, of the females, 5 segmented. While the spur count is 2-2-4, the outer spur of the front leg is very small. There is a very large pair of warts behind the eyes. Emergence of *H. borealis* occurs continuously in the summer. The other species are found in the region bordering Mexico.

13b Front tibia with 2 large spurs and no hamuli on the hind wings. 14

14a Uncommon; body not more than 6 mm in length; mesoscutum lacking warts and setae (as in Fig. 202); middle tibia often with apical spurs nearly half the length of the first tarsal segment. ...... Beraeidae

This family consists of three species of the genus *Beraea* which occur in Ontario and Georgia and questionably New York. They are about 5 mm long, the spur count is 2-2-4, the maxillary palpi are long, hairy and 5 segmented, with a short basal segment. Wing venation is reduced, the radial sector having but two branches (compare Fig. 11). The wings are about as long as the antennae.

14b Not exactly as described; apical spurs of middle tibia not more than 1/3 the length of the first tarsal segment; body size variable; mesoscutum with warts or setae. 15

15a Middle tibia without preapical spurs and with a row of black spines (Fig. 199N); with a pair of warts (w) on the mesoscutellum(s) (Figs. 199D, 211). 16

15b Middle tibia with preapical spurs; spines and warts variable. 17

16a Mesoscutum (m) with a deep, antero-mesal groove or fissure (Fig. 199D); warts located near the fissure; uncommon. .... **Sericostomatidae**

This group contains 3 North American genera and 12 species. They are found in both the east and the west but are local in distribution and rarely collected. This family once included the families now recognized as Brachycentridae, Goerinae of Limnephilidae, and Lepidostomatidae. Sericostomatidae now consists of the genera *Gumaga*, *Fattigia*, and *Agarodes*. The combination of a deep median fissure of the mesonotum, a single pair of pronotal warts, the single pair of warts on the mesoscutum near the meson distinguishes this family.

16b Mesoscutum with only a shallow antero-mesal crease (behind x, Fig. 211); mesoscutal warts some distance from the midline (arrow, Fig. 211) (See 20a). .... **Brachycentridae, in part**

17a Middle femora each with a row of 6-10 black spines on antero-ventral face (Figs. 199M, 210). .... **Molannidae**

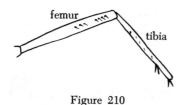

Figure 210

Figure 210 Molannidae, middle leg with spines on femur

Two genera and seven species are present. *Molannodes* is known from Alaska and the Yukon. *Molanna* is widespread in the east and is found west to Colorado and B.C. In this family the spur count is 2-4-4, with the preapi-cal spur located about 1/4 the tibial length from the apical spurs. The maxillary palpi are stout and hairy with the first two segments short, the next three longer and about equal.

17b Middle femora each with none to 2 black spines on antero-ventral face. **18**

18a Uncommon; mesoscutellum(s) with a single large oval or round wart which occupies almost the entire surface (as in Fig. 199E). **19**

18b Common; mesoscutellum (s) with two warts (w) on anterior half (Figs. 199D, 211). **20**

19a Mesoscutellum round and distinctly domelike, the wart appearing to occupy most of the sclerite; tibial spurs not hairy; maxillary palps of males 5 segmented. .... **Odontoceridae**

There are about a dozen North American species in 6 genera. Half are in the genus *Psilotreta*, and range from Quebec and New York to Tennessee and N. Carolina. Four genera representing 5 species are western, ranging from Texas and Arizona to California and Oregon. Ocelli are absent, the spur count is variable, and body length is usually about 1/2 inch.

19b Mesoscutellum triangular, only slightly convex; wart elongate and occupying mesal portions of sclerite; tibial spurs hairy; maxillary palps of males 3 segmented. .... **Goerinae, Limnephilidae**

This group is a subfamily of Limnephilidae. Ocelli are absent in the adults (present in other Limnephilidae). Five genera and about a dozen species are widely distributed but rarely collected. This group is not separated from the Limnephilidae in the larval key. Larval habitat ranges from large streams to seepage areas around springs. They feed on plant material.

20a **Middle tibia with spines, and preapical spurs of middle tibia bare and located 1/3 the distance from the apex (Fig. 199L). .... Brachycentridae**

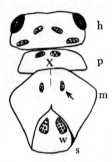

Figure 211

Figure 211 Brachycentridae, head and thorax. h—head, p—prothorax, m—mesoscutum, s—mesoscutellum, arrow—mesoscutal warts, w—warts, x—location of fissure (16a)

See also 16b. Common and widespread, this family consists of six genera and about 30 species in North America. The mesoscutellum has 2 warts as in Fig. 211. Ocelli are absent; the maxillary palps of females are 5 segmented. In males they are usually 3 segmented and pressed up against the front of the head (Fig. 199H). The spur count is variable.

20b **Middle tibia without spines; preapical spurs of tibia hairy (Fig. 199K) and located at mid-length. .. Lepidostomatidae**

Although the 70 species in this family have sometimes been divided into a number of genera, all are now considered to belong to two, *Lepidostoma* and *Theliopsyche*. The family is widespread and common. Adults have 2 warts on the mesoscutellum similar to Fig. 211. The middle tibia has spurs (which are hairy) but lacks short spines. In the female the maxillary palpi are 5 segmented but in the male the palpi appear 1 segmented and are curved up against the face (Fig. 199G). Spur count is usually 2-4-4.

21a **(4a) Spur count 3-4-4, i.e., fore tibia with a preapical spur (compare Fig. 199L) (see note). .... Polycentropodidae**

(*Note*: *Cernotina,* of this Family, lacks a preapical spur. See Ross 1944.)

In our region this family consists of 7 genera and about 70 species. Body length is usually 8-12 mm. In both this family and Psychomyiidae the 5th segment of the maxillary palp is elongate and multiarticulate, the ocelli are absent, and the middle tibia is often dilated in females. They are found near flowing water and are widespread and common.

21b **Spur count 2-4-4, i.e., fore tibia without a preapical spur (compare Fig. 199N). .. .. Psychomyiidae**

There are 5 genera and about 15 species north of Mexico. Adults are usually 4-6 mm in length. See discussion, Polycentropodidae (Couplet 21a).

COLEOPTERA
Beetles

Worldwide there are over 1/4 million described species of beetles, and about 30,000 are known for North America. One hundred twelve families are recognized in North America but fewer than a dozen are truly aquatic. About 20 families are covered in the following keys, which include semiaquatic and marine beetles. Beetles may be aquatic in the larval or adult stage, or both. Unlike the Diptera, which show extreme modifications in the larval stage, the Coleoptera have generally retained the 3 pairs of true thoracic legs and basic chewing mandibles. Thus, most either swim or crawl, and most either graze on plant material or attack prey with the mandibles. There are few sedentary or burrowing forms as are common in the Diptera.

The truly aquatic families include Gyrinidae, Haliplidae, Dytiscidae, Hydrophilidae (the latter two being the most common water beetles), Noteridae, Hydraenidae, Dryopidae, Elmidae, Psephenidae, and Amphizoidae. It will simplify the use of the keys if all families except the above are ignored. The semiaquatic and marine forms are included for completeness.

It is noteworthy that almost all aquatic beetles return to land to pupate, thus indicating the basically terrestrial nature of the Coleoptera. Adults carry an air supply under the elytra or on a velvety layer of surface hairs. Larvae respire through the integument or through gills. The flying powers of most adults insure that almost all aquatic habitats are soon colonized by beetles. Their habitats range from mountain torrents to stagnant ponds and ditches. Gyrinid adults live on the water surface. Aquatic beetles are about equally divided between predatory and herbivorous forms.

All stages can be preserved in alcohol, or adults may be mounted on pins.

COLEOPTERA LARVAE

1a Without legs (see Fig. 24 and Couplet 12a, key to Orders) (Snout Beetles) (Water Scavengers). .... **Curculionidae, Hydrophilidae**

Larvae of Curculionidae are among the few legless aquatic beetle larvae. The mouthparts are directed ventrally and the body is membranous except for the dorsal prothorax. Larvae usually burrow through plant tissue or make their way between the plant stem and leaf sheath. In general they have few aquatic adaptations. One species is able to secrete a watertight cocoon connected to the air-filled stem of aquatic plants and thus remains submerged during pupation. Another has hook-like spiracles which can pierce air-filled plant tissue for respiration. One genus of Hydrophilidae also lacks legs (see Couplet 16b, this key).

1b Legs present. 2

2a Legs 5 segmented (coxa, trochanter, femur, tibia, tarsus, plus claws) (Fig. 216B); with two tarsal claws except in Haliplidae. 3

2b Legs 4 segmented (coxa, trochanter, femur, tibiotarsus, plus claw, Fig. 217C); tarsal claws single. 8

3a Legs with 1 claw (Fig. 212) (Crawling Water Beetles). .... **Haliplidae**

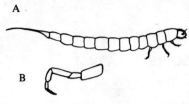

Figure 212

Figure 212   A. Haliplidae; B. leg

The combination of 5 segmented legs plus a single claw immediately separate haliplid larvae from all others. Some species have greatly elongated spines on the surface of the body (Fig. 32C). They are superficially rather similar to elmids and their relatives but the latter have 4-segmented legs. Larvae feed on plant material such as algae filaments, but some are said to feed on small aquatic worms. Pupation occurs on shore. They are usually found in still water. There are 3 North American genera and the family is common and widespread.

3b   Legs with 2 tarsal claws. 4

4a   Last (10th) abdominal segment with 4 claws (Fig. 213); elongate lateral gills present on nine abdominal segments (Whirlygig Beetles). .... Gyrinidae

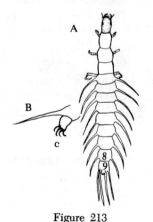

Figure 213

Figure 213   A. Gyrinidae; B. 10th abdominal segment and claws (c)

These larvae have 2 tarsal claws on each leg and there are 10 abdominal segments. They are separated from other larvae with lateral gills by the 4 hooks on abdominal segment 10. Larvae are carnivorous. They leave the water to pupate, forming a pupal cell of mud or plant material on some standing object. Overwintering is in the adult stage. Larvae are found in still water and the group is very common and widespread.

4b   Last segment of abdomen without 4 hooks. 5

**5a   Abdomen with 9 segments, sometimes with a ventrally directed tenth segment; segment 9 bearing a forked process (Fig. 214) (Ground Beetles). ....
Carabidae (including Omophronidae)**

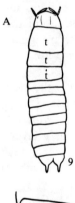

Figure 214

Figure 214 Carabidae. A. Dorsal view of larva with 9th segment; B. ventral view of 9th and 10th segments of abdomen, t—thoracic segments

These larvae are not aquatic but some species live near the water's edge where they burrow in search of insect prey. Carabidae is a very large family (20,000 world species) and only a few are associated with water. One form known from California lives on the seashore and is submerged at high tide.

5b Abdomen with 8 visible segments. 6

6a Larvae wire-worm like, similar to Fig. 220; legs short and modified for digging; a chitinous point at the end of the abdomen (Burrowing Water Beetles). ....
.... **Noteridae**

Adult Noteridae resemble Dytiscidae, but the larvae are very different in the two families. Noterid larvae are elongate burrowing forms which can dig rapidly through the substrate. They live among and feed on the roots of plants submerged under one or two feet of water. Larvae do not come to the surface for air but use the point at the end of the abdomen to pierce air-filled plant tissue. Pupation occurs attached to a plant under water, and the cocoon is filled with air obtained from a hole chewed into the plant stem or root. Winter is apparently passed in the adult stage; eggs are laid in the spring and newly emerged adults can be found in the fall. The family is quite common and widespread but often overlooked.

6b Larva not a digging form, adapted for running or swimming. 7

7a Larvae somewhat flattened and sides of abdomen expanded as plates (Fig. 215); ventral side of head with a single suture near the midline. .... Amphizoidae

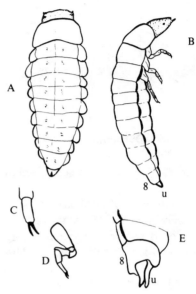

Figure 215

Figure 215 Amphizoidae. A. Dorsal view; B. lateral view; C. claws; D. leg; E. end of abdomen; u—urogomphi or cerci

Larvae are up to 15 mm long and they can be found crawling on the bottom of streams or among trash and log jams. The two tarsal claws, 5 segmented legs, 8-segmented abdomen, and flattened lateral plates make these uncommon larvae easy to identify. They are predatory and sometimes remain above the water, re-entering to search for prey. They are rarely collected.

7b Without thin lateral plates on abdomen (Fig. 216); ventral side of head with two sutures near the midline (Predaceous Water Beetles). .... Dytiscidae

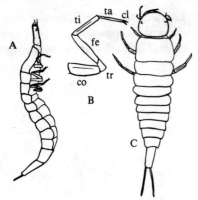

Figure 216

Figure 216 Dytiscidae. A. Lateral view; B. leg; C. dorsal view, co—coxa, tr—trochanter, fe—femur, ti—tibia, ta—tarsus, cl—claw

While quite variable in body shape, larvae of this family all have large, powerful, grooved mandibles, 8 visible abdominal segments, and 5 segmented legs with two claws. The body is "submarine" shaped. There are fringes of hair which aid in swimming on various parts of the body, and the legs are long and thin. The genus *Coptotomus* has 5 pairs of lateral gills on the abdominal segments but they are easily separated from the gill-bearing Megaloptera and Gyrinidae (and others) by characters given in the keys.

There is usually one generation per year. Adults usually are the overwintering stage. Eggs are laid in a variety of places, including incisions in plant stems. Pupation occurs on shore above the water line. Larvae catch prey with their hook-like mandibles, first injecting digestive fluid, then sucking out the fluid. Dytiscidae are very common and widespread.

8a   **Body flat and oval, head hidden from dorsal view (Fig. 217) (Water Pennies).** .... **Psephenidae**

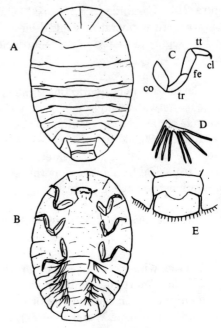

Figure 217

Figure 217 Psephenidae. A. Dorsal view; B. ventral view; C. leg; D. gill; E. detail of ventral apex of abdomen, co—coxa, tr—trochanter, fe—femur, tt—tibio tarsus, cl—claw

These larvae vary from round to pear-shaped (Figs. 217, 219A). They can be separated from all other larvae by the head which is hidden in dorsal view. Larvae of several species have not been described. They crawl over stones in rapid currents or sometimes live on pond bottoms or wave-washed shores. They feed on plant material. Some have ventral gills on the abdomen (Fig. 217B, D).

8b   **Body less flat and oval, head visible from dorsal view.** 9

9a   **Antennae about half the length of the body (Fig. 218) (Marsh Beetles).** .... **Helodidae**

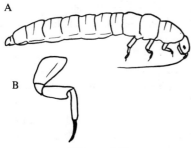

Figure 218

Figure 218 Helodidae. A. Lateral view; B. leg

Helodid larvae have 4 segmented legs with a single claw and a very long antennae, the latter very uncommon among the larvae of holometabolous insects. Adults are terrestrial and probably lay eggs in damp areas near water. There is almost no published information on the biology of North American larvae. European forms are said to feed on duckweed and algae. Larvae of some leave the water to pupate. One species is abundant in water-filled tree holes. While apparently widespread and not uncommon, little is known of this group.

9b   Antennae very short. **10**

10a   Body somewhat pear-shaped in dorsal view, thorax much wider than the head or abdomen; tip of abdomen with a ventral adhesive sucker; rare, southwest U.S. .... **Hydroscaphidae**

A single species is known from California, Arizona, and Nevada. Larvae feed on algae and live on stones in streams. They are about 1 mm long.

10b   Not pear-shaped and without a ventral sucker at end of abdomen. **11**

11a   Body a series of well sclerotized rings or segments and *usually* with an operculum (o) or lid-like structure on ventral side of ninth abdominal segment (Fig. 219B) (Riffle Beetles). .... **Dryopoid Beetles, in part, 12**

(Psephenidae are also members of this superfamily; see Couplet 8a.)

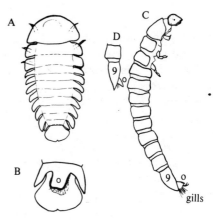

Figure 219

Figure 219 Dryopoidea. A. Psephenidae; B. ventral view of tip of abdomen of same; C. Elmidae; D. tip of abdomen of same, o—operculum, 9—ninth abdominal segment

11b   Without an operculum on ninth segment *and* body not as in Figs. 219 to 221; body often fleshy. **15**

12a   Ninth segment without an operculum. .... **Ptilodactylidae**

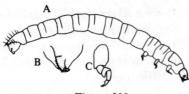

Figure 220

Figure 220   Ptilodactylidae. A. Lateral view; B. detail of ventral apical abdomen, lateral view; C. thoracic leg

Larvae live in springs or streams where they probably feed on plant material. Three species in 3 genera, 1 eastern and 2 western, are known to be aquatic. These larvae are almost round in cross section. They differ from other riffle beetles in lacking an operculum on segment 9. One genus has ventral gills. The other two have curved anal legs covered with short spines (Fig. 220B). The body may exceed 25 mm in length. They are seldom collected.

12b   **Ninth segment with an operculum and caudal chamber (Fig. 219B, O).** 13

13a   **Body cylindrical, circular in cross section; segments of the abdomen are in the form of almost complete rings on the first 5 segments; the rings are complete on segments 6-9; without gills in caudal chamber (Fig. 221) (Riffle Beetles).** .... **Dryopidae**

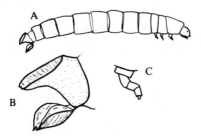

Figure 221

Figure 221   Dryopidae. A. Larva, lateral view; B. caudal chamber, lateral view; C. leg

Retractile gills are absent in the caudal chamber (Fig. 221B) and there are lateral spiracles on abdominal segments 1-7. The genus *Helichus* has about 10 aquatic species. In this group the operculum has 2 toothlike tubercles on the posterior margin, the ninth segment is flattened dorsally, and it has a notch at apex similar to Fig. 221B. These beetles are found on debris, under rocks, or on wood in streams throughout North America. They are seldom collected.

13b   **Body somewhat flat on the ventral side, anterior segments of the abdomen are not complete rings, but with distinct sternites ventrally; with retractile filamentous gills in caudal chamber (Fig. 219C).** 14

14a   **Apex of ninth abdominal segment notched; abdomen with ventral plates on the first 6 segments (Fig. 219C, D) (Riffle Beetles).** .... **Elmidae**

These beetles are common and widespread. There are over 2 dozen genera and about 100 species in North America. Larvae feed on plant material and can be found on wood, in gravel, or on rocks in streams throughout North America.

14b   **Apex of ninth abdominal segment rounded; ventral plates on first 4 segments of abdomen only.** .... **Limnichidae**

Larvae range from aquatic to terrestrial, and probably less than half a dozen species are aquatic. One group is intertidal. They are seldom collected but are widespread. The only described larvae belong to the genus *Lutrochus*. The body is more robust than in Elmidae, and the head is nearly as wide as the thorax.

They are usually found on rocks or wood in streams.

15a **True aquatics, swimming or on aquatic plants; abdomen lightly sclerotized, fleshy and wrinkled in appearance. 16**

15b **Shore or intertidal insects, well sclerotized plates and setae are usually visible on abdomen. ....**

A number of beetle larvae are semi-aquatic or intertidal, and among those most likely to be encountered are the following:

Larvae of *Hydraenidae* (=Limnebiidae) are terrestrial carnivores living at the water's edge and are characterized by 4 segmented legs, a single tarsal claw, the tenth abdominal segment usually with a pair of recurved claws, and a pair of cerci at the end of the abdomen. The mandibles have basal molar tubercles. Adults are aquatic.

Staphylinidae, the Rove Beetles, include a number of tiny intertidal species and some which occur on freshwater shores. Larvae have 4 tarsal segments, a single claw, segmented cerci, ocelli in groups of 5, and mandibles which are in the shape of a smooth sickle-like blade and lack a basal molar area; claws on the tenth abdominal segment are lacking.

Heteroceridae are elongate and slender and have 4 leg segments with a single tarsal claw; cerci are absent, gills are lacking, the antennae are very short, there are 5 ocelli on each side, and there are 10 abdominal segments, the tenth small and ventral with a two-jointed retractile prolongation. The mandibles have a ventral crushing tubercle. They live in wet sand or mud along shores.

Eurystethidae are intertidal and found on the west coast from California to Alaska. The abdomen terminates with 4 prongs, two directed inward and two outward and upward.

Melyridae occur on the seashore from British Columbia to California. The abdomen ends with two sclerotized prongs and sometimes a pair of smaller teeth between; the outer pair are not curved as in the previous family.

16a **Legs short and poorly developed; mouthparts ventral; feeding on submerged or floating plants (Fig. 222) (Leaf Beetles). .... Chrysomelidae**

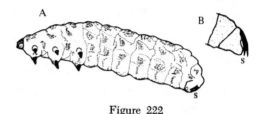

Figure 222

Figure 222 A. Chrysomelidae, *Donacia*; B. detail of spiracles, s—spiracular spines

Larvae of the subfamilies Donaciinae and Galerucinae of this primarily terrestrial family feed on the leaves of water lilies, burreeds, *Myriophyllum* and arrowheads (*Sagittaria*). Many of the beetles are host specific. All stages of *Galerucella nymphaeae* (Galerucinae) feed on the upper surface of leaves of water lilies. Larvae of the latter cannot swim and no life stage shows special adaptations to aquatic life. Larvae of the genera of Donaciinae (about 60 species) have ventrally directed elongate and spinelike spiracles (s) on abdominal segment 8 with which air-filled plant stems are pierced for respiration. They feed on submerged plant parts. Larvae spin an air-filled cocoon which is attached under water to the host plant.

16b **Legs longer and well developed; mouthparts directed anteriorly; free swimming and predatory (Fig. 223) (Water Scavenger Beetles). .... Hydrophilidae**

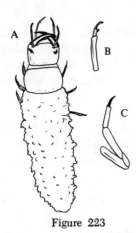

Figure 223

Figure 223 A. Hydrophilidae; B. detail of maxilla; C. leg

One small group of this family is terrestrial but most are aquatic. Larvae are predatory and even cannibalistic. Some species reach more than 2 cm in length. Larvae are usually found in shallow and quiet waters. Most larvae have well developed mandibles and legs, but *Cercyon* sp. lacks legs (Order key, Couplet 12a). The abdomen is largely membranous and wrinkled, sometimes with long lateral filaments similar to Fig. 213. This is one of the largest and most common families of aquatic beetles.

COLEOPTERA ADULTS

1a **Eyes divided so that head has four large round eyes, two visible from above and two visible from below (Fig. 224) (Whirlygig Beetles). .... Gyrinidae**

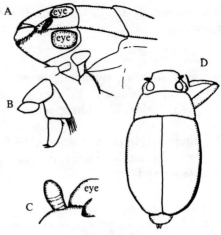

Figure 224

Figure 224 Gyrinidae. A. Side view of head with upper and lower eyes; B. hind leg; C. antenna; D. dorsal view, whole insect

These beetles are blackish, oval, somewhat flattened, often shiny, and 6 to 10 mm long. They inhabit the water surface film but are capable of diving if necessary, using an air supply that is carried under the elytra. The antennae are short and thick (Fig. 224C). The middle and hind legs are short and flattened and do not reach beyond edge of body. The front legs are longer and adapted for grasping prey which consists of insects that have fallen onto the water surface. In the field, the peculiar circling, gyrating, and whirling manner of swimming is unmistakable and gives them their common name. About 60 species and three genera are found in North America. Adults of *Gyretes* and *Dineutus* (scutellum not visible) are typically stream inhabitants while *Gyrinus* (with a visible scutellum) is associated with lakes and ponds. Many species fly well and colonize any habitat large enough to swim on. The beetles often form large multi-species schools on the water surface. The family is widespread and common. The adults are more often seen than the larvae. Eggs are attached to submerged plants, larvae are strictly aquatic, and pupation occurs on land. Pupae are attacked by numerous species of parasitic Hymenoptera.

1b **Eyes not divided; head with two eyes only.** 2

2a **Hind legs projecting from beneath a large plate (P) which covers much of the abdomen (Fig. 225) (Crawling Water Beetles).** .... Haliplidae

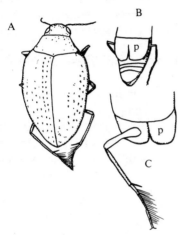

Figure 225

Figure 225 Haliplidae. A. Dorsal view, whole insect; B. ventral view of plate and hind legs; C. dorsal view of coxal plate and leg, abdomen removed

These beetles are small (1.5-4.5 mm), oval, hump-backed, brown to tan, and they usually have black spots on the elytra. They are poor swimmers but crawl over the substrate in ponds, lakes, or quiet streams. Antennae are 11 segmented. An air supply is carried under the elytra and under the enlarged hind coxal plates. This is both an air supply and a hydrostatic organ. There is disagreement about the food habits of adult haliplids. Some reports say they are predatory on other insects and microorganisms while others indicate they

feed primarily on the filamentous algae with which they are often associated. Haliplids are widespread. There are two common genera, *Peltodytes* with the coxal plate reaching the last abdominal segment, and *Haliplus* with the plate reaching the third abdominal segment.

2b **Ventral abdomen not covered by large plates.** 3

3a **Ventrally, first abdominal segment (1) completely cut in two by attachment of hind legs (Fig. 226); antenna never clubbed.** 4

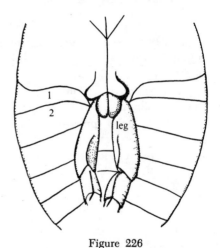

Figure 226

Figure 226 Ventral view of thorax, abdomen, and rear legs, with segment 1 divided by the coxae (attachment of hind legs)

3b **First abdominal segment (1) not divided by the attachment of the hind legs (Fig. 227); antenna sometimes clubbed.** 8

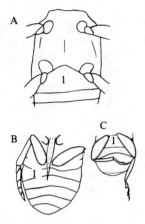

Figure 227

Figure 227 Ventral view of three beetles, segment 1 undivided by the coxae

4a   **Primarily terrestrial, intertidal or shore insects; with at least some individual erect setae or hairs on various parts of the body (ventral abdomen, femora, ventral mouthparts); without fringes of swimming hairs on legs (Fig. 228) (Ground Beetles).** ....
.... **Carabidae, including Omophronidae**

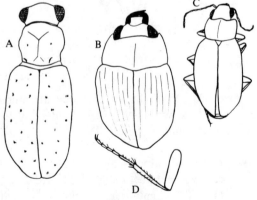

Figure 228

Figure 228   Carabidae. A. *Elaphrus;* B. *Omophrons;* C. "typical" Carabid; D. rear leg of B

(Amphizoidae, Couplet 5a, may be confused with the group.) Although primarily a terrestrial Family, several genera are associated with the water's edge and are therefore included. The genus *Elaphrus* (Fig. 228A) is usually mistaken at first glance for a tiger beetle (Cicindellidae). One half inch or less in length, *Elaphrus* is characterized by the round protruding eyes, the head which is as broad as the prothorax, and by the 3 rows of indentations as well as additional shallow punctuations on each elytron. These are a predominantly northern group with 14 species in North America. They are widespread and all are predators which are found near water, either among dead leaves and vegetation or on wet soil. Also of interest is *Omophron* (Fig. 228B) subfamily Omophroninae, which is sometimes given family status. These are circular in outline; the legs (Fig. 228D) and antennae are long. The scutellum is not visible and there are 14 or 15 longitudinal striations on the elytra. The coloring is often metallic over a pale background. There are 11 North American species and while widespread, are seldom collected. They are active at night but can be flooded from their burrows in the bare soil and sand near water. Fig. 228C shows a "typical" carabid; various species and genera with this appearance can be found near water.

4b   **Aquatic beetles (but occasionally collected at lights or in flight); body without individual erect hairs or setae; a fine pubescence or fringes of swimming hairs on legs (Fig. 231B) is often present.** 5

5a   **Hind legs without swimming hairs; uncommon western beetles, body not smoothly oval (Fig. 229).** ....
.... **Amphizoidae**

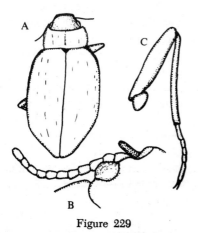

Figure 229

Figure 229 Amphizoidae. A. Body from dorsal view; B. eye and antenna; C. rear leg

In this group the tarsi are 5 segmented and the scutellum is visible. These are slow crawling beetles about 10-15 mm in length which live on stones or wood and trash in streams or rivers. While related to Carabidae, they resemble Tenebrionidae. They live under water but the legs are not modified for swimming. An air supply is carried under the elytra. Eggs are laid in late summer and larvae emerge in the fall. The life cycle has not been observed in detail but larvae may require several years to develop and pupation apparently occurs away from the water. There are 4 species in the single genus *Amphizoa* ranging from Alaska to California and inland to Alberta and Montana. One other species occurs in Tibet.

5b   Hind legs with swimming hairs; body streamlined and oval; common. 6

6a   Front and middle tarsi appearing 4 segmented, the third segment deeply bilobed (Fig. 230C); scutellum almost never visible, (see Couplet 7b). .... **Dytiscidae in part**

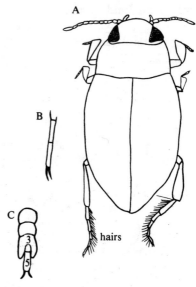

Figure 230

Figure 230 Dytiscidae, Hydroporinae. A. Whole insect, dorsal view; B. rear tarsus and claws; C. front tarsus

6b   Front and middle tarsi clearly 5 segmented; scutellum variable. 7

7a   Hind tarsi with two claws of equal length; scutellum not visible (as in Fig. 230) (Burrowing Water Beetles). .... **Noteridae**

These beetles are usually under 6 mm in length. They resemble Dytiscids but are separated by the two equal claws of the hind tarsi, the 5 segments of equal length in the front and middle tarsi, and the scutellum which is not visible. Adults are strong swimmers but both adults and larvae often burrow in the bottom of plant-filled ponds or lakes and occasionally rivers. While widespread and fairly common in North America, most of the information of this family is based on European studies.

7b Either scutellum is visible, or if not visible, then hind tarsi each with a single claw (Figs. 230, 231) (Predaceous Diving Beetles) (see also Couplet 6a). .... **Dytiscidae**

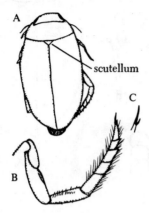

Figure 231

Figure 231 Dytiscidae. A. Body from above; B. hind leg; C. detail of claws

These are oval beetles with long antennae and are the common "water beetles" known to almost everyone. They are rather similar in shape and habitat to the Hydrophilidae but are very distantly related. The antennae of Hydrophilidae are clubbed; antennae are filiform in Dytiscidae. Both are common and abundant and they are the 2 major aquatic families. Dytiscid adults are predators. The body is oval from lateral view and they move their legs in unison for efficient swimming. In contrast, the hydrophilids are true to their name of water scavenger, and they are poor swimmers which move their legs alternately in a crawling motion. To obtain air, hydrophilids break the water surface with the antennae and "shoulder", while dytiscids break the water surface with the tip of the abdomen. Adults of both often fly to lights at night. Dytiscidae are very common and widespread and there are many species. They range in size from tiny to over an inch in length. Dytiscid beetles (as well as

other families) are often attacked by ascomycete fungi of the family Laboulbeniales as well as by parasitic wasps.

8a Minute, rare or locally abundant beetles under 1 mm in length with hind tarsi 3 segmented; known from the western and southwestern U.S. **9**

8b Small or large beetles with more than 3 segments in hind tarsi. **10**

9a Elytra short (similar to Fig. 234) with conical abdomen exposed; antennae 8-segmented (see also Couplets 12a and 15a) (Skiff Beetles). .... **Hydroscaphidae**

These beetles are about 1 mm long and live on filamentous algae growing on rocks in streams. The elytra are short with fringe hairs of the wings projecting from beneath. The general appearance is a small red-brown Staphylinid. There is a single species in North America, *Hydroscapha natans,* known from California, Nevada, and Arizona. They sometimes occur in populous colonies, adults living with the larvae. Habitat ranges from icy streams to hot springs.

9b Elytra not short; oval, convex beetles; antennae 11 segmented. ...... **Sphaeriidae**

There are 2 or 3 species known from California, Texas and Washington where they are found on mud or under stones near water. The larvae are unknown. There is a single genus, *Sphaerius.*

10a Antennae shorter than the maxillary palpi; antenna terminating in a 3-5 segmented club which is set off by an enlarged or cup-like segment (Figs. 232, 233). **11**

Note that the antennae of Dryopidae form a terminal pectinate club of different structure than described; the last tarsal segment in this family is as long as the preceding 4 combined. Heteroceridae have 7 enlarged apical antennal segments and have flat spiny tibiae. (In these cases go to 10b.)

10b Antennae longer than the maxillary palps; club absent or not as described. 12

11a Club of antennae with 5 segments (Fig. 232C); 6 or 7 sternites visible, body rarely exceeding 1.7 mm in length. .... Hydraenidae (= Limnebiidae)

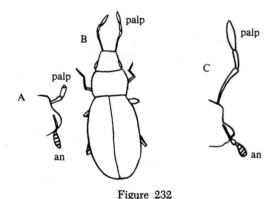

Figure 232

Figure 232 Hydraenidae. A. Antennae and palps as in *Ochthebius* and *Limnebius*; B. body as in *Hydraena*, dorsal view; C. detail of palps and antennae as in *Hydraena*. an—antenna

While early studies placed these in the Hydrophilidae, it is now agreed that the resemblance is superficial and they are actually more closely related to the Staphylinidae. They can be separated from Hydrophilidae by the 5 segmented antenna club and six or seven rather than 5 visible abdominal sternites. Larvae live on shore, swim poorly, and resemble those of Staphylinidae. These beetles are small,

1 to 3 mm in length, and cling to submerged objects in streams or creep over the bottom of ponds, sometimes burrowing in the sand or mud. Some are intertidal. Adults feed on plant material in the water but larvae are predatory shore dwellers. There are three North American genera. *Ochthebius* has many species and is recognized by a thin transparent edge around the sides of the pronotum as well as metallic colors. The palps are short (Fig. 232A). *Hydraena* have greatly elongated maxillary palps (Figs. 232B and 232C) and coarsely punctuate elytra. *Limnebius* have short palps as in Fig. 232A but lack the thin pronotal edge of *Ochthebius*. *Ochthebius* is most common in standing water while the other two genera frequent streams and rivers. Adults carry a bubble of air on the ventral side of the body which is renewed by crawling to the surface. Eggs are laid in the spring, sometimes protected by silk, either in or out of the water. Larvae mature in about two months. Pupation occurs on shore and adults are the overwintering stage. Adults are known from hot spring and saline pools as well as fresh water.

11b Club of antenna with 3 segments plus a cup-like segment; abdomen with 5 sternites visible (Fig. 233C) (Water Scavenger Beetles). .... Hydrophilidae

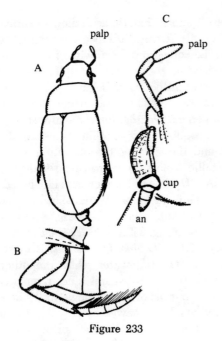

Figure 233

Figure 233 Hydrophilidae. A. Body from dorsal view; B. hind leg; C. head from ventral view. an—antenna

This is a common and easily recognized group of beetles. Some authors raise certain subfamilies of this group to family status, for example, Hydrochidae and Helophoridae, but they are not separated in this key. One group of hydrophilids (subfamily Sphaeridinae) is terrestrial, living in dung or moist earth. The aquatic species are among the most common, abundant and widespread aquatic beetles (along with Dytiscidae) being found virtually everywhere. The maxillary palps are long and easily mistaken for antennae at first glance (Fig. 233A). The antennae are comprised in total of 7 to 9 segments with the last 3 clubbed, the club being set off by an enlarged segment (cup, Fig. 233C), and the entire structure is folded under the head. The beetles are 2-40 mm in length, typically shiny and black or brown with a convex appearance, sometimes with punctations or longitudinal striations. They often have a sharp spine projecting between the hind legs

(subfamily Hydrophilinae). Compared to Dytiscidae they are flat on the underside. The legs move alternately while swimming in the manner of a walking insect; air is picked up by breaking the surface with the antennae and front corner of the prothorax. They are sometimes called "silver beetles" because of the shining film of air on the under surface of the body. Adults feed primarily on plant material but a few are predatory. Adults fly well and they are often seen in large numbers at lights during dispersal flights. Eggs in some species are laid in silk cases, and a few carry the cases under the abdomen. Larvae are most commonly found in summer. Pupation occurs in damp soil on shore. Adults prefer weedy shallow ponds but can be found in flowing water.

12a   Elytra short, exposing a large portion of the dorsal abdomen, and wings present (Fig. 234) (Rove Beetles). ....
... Staphylinidae

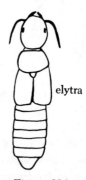

Figure 234

Figure 234 Staphylinidae. Dorsal view of whole insect

These are not truly aquatic but many species live in vegetation or burrow in sand or mud at the edges of aquatic habitats. Some are found at the seashore and are active under water on the bottoms of tide pools. Adults are predatory.

(Note: also with short elytra, but not related to Staphylinidae, are species of the genus *Endeodes* of the family Melyridae which are intertidal, from British Columbia to California. (See also Couplets 15a and 9a.)

12b   Elytra longer, covering all of the abdomen, *or* if elytra are short, then hind wings are absent. 13

13a   Front of head beyond eyes prolonged into a beak or snout (Fig. 235) (Snout Beetles). .... Curculionidae

Figure 235

Figure 235 Curculionidae. Lateral view of thorax and head with beak

This family is unmistakable because of the elongated beak. The family is very large (over 3000 species in North America) and both adults and larvae feed on plant material. About 100 species are known to feed on aquatic plants. Some adults spend long periods of time on the undersides of floating leaves and some have been observed mating under water or walking on pond bottoms. No specific aquatic adaptations are present in the adults. One species is a pest of cultivated rice. Aquatic snout beetles are probably widespread but they have been studied in detail in only a few cases and they are usually overlooked.

13b   Front of head not elongated into a snout. 14

14a   Marine or intertidal insects with base of trochanter not covered by a projection from the hind coxa; range, California to Alaska. 15

14b   In or near freshwater, rarely marine; base of trochanter covered by a projection or lobe from the hind coxa (Fig. 236B). 16

15a   Elytra short, wings absent; all tarsi 5 segmented (see also Couplet 12a). .... Melyridae

Most members of this family (the soft-winged flower beetles) are terrestrial. They have soft elytra which are widest near the tips. They are usually found on flowers where they feed on soft-bodied insects or insect eggs. One genus (*Endeodes*), with about half a dozen species, occurs on ocean beaches from B.C. to California. These small predatory beetles live in debris or on rocks near the high tide mark and are often covered by water. They have peculiar retractile membranous vesicles between the thoracic segments.

15b   Elytra not short, hind tarsi 4 segmented. .... Eurystethidae (=Elacatidae=Salpingidae)

Several species of the genus *Eurystethes* live in the intertidal area and range from California to Alaska. They are probably predatory, live in colonies, and are 2-3 mm long. They are flattened and have widely separated coxae. They are black or metallic green and have sparse short hairs on the body.

16a Tarsi appearing to have 4 segments, segment 4 minute, concealed in bilobed segment 3 (Fig. 236A) (Leaf Beetles). .... Chrysomelidae

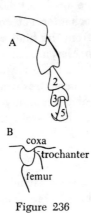

Figure 236

Figure 236 Chrysomelidae. A. Tarsal segments; B. base of rear leg

This is a large family of primarily terrestrial beetles, and only a few are aquatic. Potato beetles are a typical member of the group. The family is characterized by an oval or elongate shape, long antennae, and tarsi with a short segment 4 hidden between 3 and 5. Aquatic members show no obvious aquatic adaptations. They are usually coppery, green, or black in color and can be collected by sweeping aquatic vegetation or skimming the water surface. The adults usually feed on the surfaces of floating leaves or go to the water's edge to lay eggs, to only occasionally venturing under water. Aquatic Chrysomelids belong to the subfamilies Donaciinae and Galerucinae. In the latter there is the single aquatic species *Gallericella nymphaeae*, which lives on water lilies. There are about 60 species and two genera in Donaciinae. *Neohaemonia*, with a single species, is usually found on the plant *Potamogeton natans*. *Donacia*, with numerous species, is found on a variety of plants. Aquatic leaf beetles are common and widespread but often overlooked.

16b Tarsi not with segment 4 minute and hidden in bilobed segment 3. 17

17a Small oval beetles, 3 to 4 mm long; black to yellow-brown in color with scutellum visible; tarsi 5 segmented, antennae 11 segmented. .... Helodidae

While larvae are aquatic, adults are terrestrial. Adults of *Elodes* are able to swim but return to dry land at the first opportunity. They are not rare but are seldom collected.

17b Not as described above. 18

18a Tarsi with 5 segments; tibiae; not flattened and without spines (Fig. 237) (Riffle Beetles). .... Superfamily Dryopoidea, 19

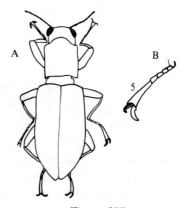

Figure 237

Figure 237 Elmidae. A. Whole insect, dorsal view; B. tarsus with elongate 5th segment

Elmidae is the most commonly collected family of this group. Other families are separated in the following keys.

18b Tarsi with 4 segments; front and middle tibiae greatly flattened and expanded and with spines on the outer edge (Variegated Mud-loving Beetles). .... Heteroceridae

These beetles are flat, rectangular and nearly twice as long as wide, and are covered with a dense pubescence. There is a single genus, *Heterocerus*. The apical 7 segments of the antennae are thicker than the first 4. Mandibles are long and project in front of the head. The body is black or brown with wavy bands or spots of yellow. They burrow in mud or sand along shores of streams or lakes and can be collected by splashing water onto the shore to drive them from their holes. They are widespread but not often collected.

19a   Apical tarsal segment not longer than the basal 4 combined; body oval, convex, clothed with dense short hairs. .... Limnichidae

The legs are retractile, i.e. can be drawn into grooves, the middle coxae are widely separated, and the hind coxae are close together. Adults often occur at or above the water line on rocks in streams, on the underside of submerged rocks, or in burrows on the shore. They can be flooded out for collecting. One genus is intertidal. They are widely distributed but not commonly collected (see Couplet 10a).

19b   Apical (5th) tarsal segment longer than the basal 4 combined (Fig. 237B); body not exactly as described. 20

20a   Elytra and body soft and leathery. 21

20b   Body hard and well sclerotized. 22

21a   Body round to oval in dorsal view, length less than twice the width; mandibles and labrum usually not visible from above because front of head is produced for-

ward as a shelf between the antennae (Water Pennies). .... Psephenidae

Adults are usually found along the banks or on rocks protruding from streams. There are about a dozen species and they are widespread, being found wherever there is suitable flowing water, or in some cases, wave-washed shores. They are more diverse in the west. *Psephenus lecontei* is the common eastern species. Body length 6 mm; antennae in males comb-like.

21b   Body elongate, length about 2 1/2 times the width; labrum and mandibles usually visible. .... Ptilodactylidae

Adults are found on vegetation near the streams or springs inhabited by the larvae. They are uncommon, known from California, Oregon and Nevada, and from Georgia to New York. There are three species in 3 genera which are considered to be aquatic. *Stenocolus* may be over 20 mm in length; the others are from 5-10 mm long.

22a   Antennae slender (Fig. 237); anterior coxa globular. .... Elmidae

These beetles are usually under 4 mm in length and are very common and widespread. They are usually found in streams, either under water on the substrate or on rocks just above the water line. While they can be considered as truly aquatic, they are not able to swim, instead crawling over the surface as do other Dryopoidea. Air is carried under the elytra as is usual in Coleoptera and also in a velvety layer of hairs over the body. Pupation occurs above the water line. Adults feed on plant material. These are the most common and widespread dryopoid beetles, with about 25 genera and nearly 100 species.

22b Antenna forming a pectinate club; anterior coxa transverse. .... Dryopidae

These beetles are 4-8 mm in length and cylindrical and elongate in shape. The head can be partly withdrawn into the thorax, leaving only the eyes and a small part of the head exposed. Females have a sharp ovipositor. Some inhabit debris in streams just above the water level, some live under rocks in streams, and a few are found on aquatic plants or debris in ponds or swamps. Adults are often attracted to lights.

Air is carried under the elytra, and much of the body is covered with a velvety pile of hairs which contains a silvery layer of air. Adults feed on plant material. There are three genera and about a dozen species. They are widely distributed but not often collected.

DIPTERA
True Flies

Many Diptera are aquatic in the larval stage. Some adults "skate" on the water surface or are restricted to shores and beaches. Diptera is one of the large Orders of insects with about 90,000 known species worldwide and about 17,000 species in North America (north of Mexico). One hundred seven families are recognized in North America and 23 families are covered in these keys. Of these, less than a dozen families are aquatic in the narrow sense, with larvae of all species developing in water. These are mostly mosquito and midge-like primitive flies (Nematocera). Other families may contain several species which are truly aquatic while the majority are terrestrial, or the members may be semi-aquatic. The diversity of habitats, habits, and appearance in the Diptera is so great that few generalizations can be made. Some larvae inhabit extremely polluted water while others are restricted to cool clean streams. Food gathering techniques include filter feeding, scraping and grazing on rock surfaces, and predatory behaviour. Groups such as the Chironomidae, while uniform in appearance, cover in themselves the entire range of feeding strategies from the construction of filter nets to simple grazing to active predation.

Adults of many (such as mosquitoes, blackflies, and biting midges) take a blood meal and are notorious pests. Some adults are predatory on other insects, while others probably do not feed at all. Adults have a fairly long life span and they often have sophisticated mating behavior. This usually involves swarming over a common marker and sometimes includes the offering of some prize such as a captured insect to a prospective female.

Adults can always be recognized by the lack of hind wings (with halteres replacing them). Larvae are not easily characterized as a group. Adults are usually collected, pinned and mounted dry. However, both stages may be preserved in alcohol. In mosquitoes the scales are necessary for species level taxonomy and adults should be handled very carefully and mounted on pins.

DIPTERA LARVAE

1a   Larva with a well sclerotized and easily visible head capsule (Figs. 240, 241, 242); mandibles biting and move toward each other when closing (Fig. 238) (Nematocera). 2

(Note: Stratiomyidae have a visible head capsule *and* hook-like mandibles. Select Couplet 1a in this case.)

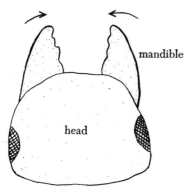

Figure 238

Figure 238 Head and mandibles of lower Diptera

1b Larva appearing headless; head capsule reduced to rods and hooks and withdrawn into the thorax (Figs. 256-258); mandibles function as a pair of parallel hooks which reach forward and rake toward the body (Fig. 239) (higher Diptera). 16

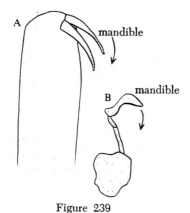

Figure 239

Figure 239 Head region and mouthparts of higher Diptera, two examples

2a With a row of six sucker-like areas on underside of body (Fig. 240) (Net-wing Midges). .... **Blephariceridae**

Figure 240

Figure 240 Blephariceridae (arrow—ventral sucker)

Larvae of this group are flat on the ventral side and appear to have 6 body regions, each with lateral lobes and a mid-ventral sucker. The head is well sclerotized. Larvae live attached to rocks in swiftly flowing water where they graze on algae. They are very common in some mountainous regions, but in general are uncommon.

2b Larvae without median sucker-like areas on underside of body and appearance not as Fig. 240. 3

3a Abdomen with 7 pairs of large lateral false legs (prolegs); antennae forked (Fig. 241) (Mountain Midges). .... **Deuterophlebiidae**

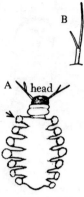

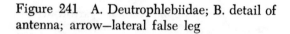

Figure 241

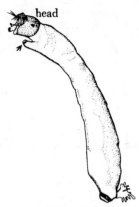

Figure 242

Figure 241 A. Deutrophlebiidae; B. detail of antenna; arrow—lateral false leg

These larvae are but two or 3 mm long and are very distinctive in shape. They inhabit mountain streams in the west where they live on smooth stones in riffles, either in splash areas or where the water is shallow and swift. Larvae have long antennae with a branch at mid-length, a 3 segmented legless thorax, and 7 pairs of lateral hooklet bearing lobes on the abdominal segments. They often occur in company with larvae of Blephariceridae and *Maruina* (Psychodidae).

3b   Abdomen without 7 pairs of lateral false legs (prolegs) and without a branched antenna. 4

4a   Posterior of abdomen enlarged so body appears club-shaped; a proleg located ventrally behind head (arrow, Fig. 242); posterior end of body with gills and a sucker-like disc and a circle of hooklets (Fig. 242) (Blackflies). .... Simuliidae

Figure 242   Simuliidae (arrow—false leg)

Larvae live in running water and sometimes thousands of individuals cover stones in riffles or rapids. Food consists of material filtered from the current by peculiar cephalic fans which· sweep toward the mouth. There is a ring of hooklets on the end of the abdomen which anchor larvae to the substrate. Silk threads are produced and sometimes larvae move downstream on the end of a thread. Blackflies are common and widespread wherever there is flowing water.

4b   Body not club-shaped; false legs and tip of abdomen not as described. 5

5a   With two pairs of prolegs behind head (H) on abdominal segments 1 and 2; tip of abdomen with two hair fringed lobes and a tube-like process (Fig. 243) (Dixid Midges). .... Dixidae

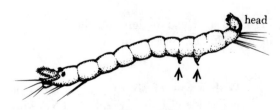

Figure 243

Figure 243 Dixidae (arrows—false legs)

Larvae are associated with the underside of the surface film in still water but may be found in quiet areas of streams. Pairs of dorsal prolegs on each of the first and second abdominal segments are diagnostic of this family. Larvae hold themselves in a U shape when at rest and the sideways "inchworm" type of locomotion of live specimens identifies them immediately. They apparently feed on organic material and are common and widespread.

5b **Prolegs and tip of abdomen not as described.** 6

6a **Larva with 3 pairs of clawed prolegs behind head and a long breathing tube (Fig. 244B) (sometimes retracted Fig. 244A) at the end of the abdomen (Phantom Craneflies).** .... Ptychopteridae

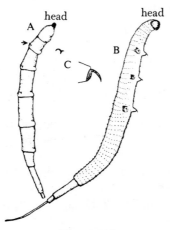

Figure 244

Figure 244 Ptychopteridae; A. *Ptychoptera* with claws (arrow); B. *Bittacomorpha;* C. detail of claw

In this family the body is elongate and slender and there is a partially or completely retractile caudal breathing tube. The integument varies

from leathery to semi-transparent and has tiny hairs (*Ptychoptera*), warty protuberances (*Bittacomorpha*) or elongate tubercles (*Bittacomorphella*). The false legs of segments 1-3 each bear a curved claw. Larvae live in mud or decaying organic material in swamps or at pond or stream margins. They are said to feed on decaying organic material. Body length, 3 cm or more.

6b **Larva without 3 pairs of clawed prolegs and a long breathing tube.** 7

7a **Thorax distinctly enlarged, segments fused and broader than abdomen; false legs absent (Figs. 245, 246).** 8

7b **Either thoracic region not distinctly enlarged compared to the abdomen *or* false legs present on the thorax or abdomen.** 9

8a **Antennae with long and strong apical spines (Fig. 245) (Phantom Midges).** .... Chaoboridae

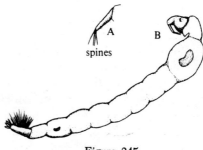

Figure 245

Figure 245 Chaoboridae; A. Detail of antennae; B. body shape

Body shape in this family ranges from that shown in Fig. 245 to very similar to Fig. 246. The apical antennal spines used in capturing prey separate this family from the true mosquitoes. Larvae are predatory on other insects, especially mosquitoes, and on planktonic crustacea. Larvae of *Chaoborus* (Fig. 245) are transparent in life with a silver air chamber at both ends. They turn white in preservative. *Chaoborus* often live a planktonic existence in large deep lakes but are also found in small permanent ponds. They usually overwinter in the larval stage and can be collected through holes in the ice.

8b **Antennae with only short hair-like bristles apically (Fig. 246) (Mosquitoes)...**
.... **Culicidae**

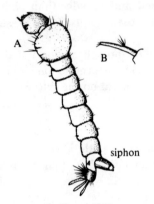

Figure 246

Figure 246   A. Culicidae; B. detail of antenna

Mosquito larvae are the familiar "wrigglers" which can be seen in standing water of all types. They rise to the surface to breathe through a sclerotized siphon (siphon reduced in some genera). They feed on surface debris or may return to the bottom substrate where they feed on organic material. Some species prefer special habitats such as tree holes or the water in pitcher plants. Others have a modified respiratory siphon and pierce the stems of aquatic plants to get air. The swollen thoracic area and lack of prolegs usually make this group easily recognizable.

9a **Head capsule covered with a chalky substance and partly withdrawn into the thorax (Figs. 247, 248) *or* head capsule completely withdrawn into thorax and usually not visible in preserved specimens without dissection (Figs. 249, 255).**
.... **10**

Figure 247

Figure 247   Stratiomyidae with head capsule partly withdrawn into thorax

9b **Head capsule completely visible, not withdrawn into thorax.** **11**

10a **Larva with about half of head capsule visible (Fig. 247); entire body usually impregnated with a chalky substance; a tuft of hairs at the tip of the abdomen, ventral hooks may be present (Fig. 248) (Soldier Flies).** .... **Stratiomyiidae**

Figure 248

Figure 248   Stratiomyidae

Larval habitat in this family ranges from damp earth and decaying vegetation to a totally aquatic existence. All aquatic species are characterized by rough, chalky, calcified skin. Larvae are somewhat dorsoventrally flattened and prolegs are absent, although ventral hooks may be present. Larvae feed on algae and dead organic material. There are many aquatic species and they are common and widespread.

10b   Head capsule usually completely withdrawn into thorax in preserved specimens; head capsule incomplete in posterior and ventral regions, but mandibles as in Fig. 238; tip of abdomen often with spiracles which are surrounded by various lobes (Figs. 249, 255) (Crane Flies). .... Tipulidae

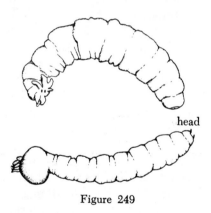

head

Figure 249

Figure 249   Tipulidae

Larvae are characterized by having a partly sclerotized head capsule which can be retracted from view (Fig. 255). Thus, they usually appear headless, but they are not. Larvae of most Tipulidae live in damp soil, moss, rotten wood, decaying leaves, or other terrestrial conditions. Others live in wet soil at the margin of water bodies while a few are strictly aquatic. Some are predatory on other insect

larvae; others feed on the roots of grasses. Many feed on plant remains and algae. For a book-length treatment of the group see Alexander (1920, Diptera references).

11a   Larva small, under 4 mm, with each body segment subdivided into 2 or 3 annuli or subsegments, these annuli often with sclerotized plates; tip of abdomen usually with a sclerotized tube and tuft of hair (Fig. 250) (Moth Flies and Sand Flies). .... Psychodidae

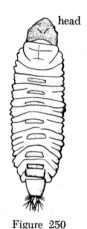

head

Figure 250

Figure 250   Psychodidae, *Maruina*

Larvae lack prolegs and the head is well sclerotized with opposed mandibles. Many genera are more elongate than the *Maruina* illustrated, resembling Fig. 251, but with characters described here. Larvae of this family range from fully aquatic to fully terrestrial. Relatively little work has been done on North American species; larvae presumably feed on organic material. *Phlebotomus*, important tropical disease vectors, live in moist areas but are not aquatic. *Maruina*, from western North America, live on stones in fast streams at or above the water line and have mid-ventral sucker disks. Others live in mud, moss, or at stream margins.

11b Body segments not subdivided into annuli, size variable. **12**

12a Prolegs (p) present on the prothorax (Figs. 251, 252). **13**

12b Prolegs absent on the prothorax. **15**

13a Dorsal body surface with fleshy tubercles and short spines or setae. .... Ceratopogonidae, in part (*Forcipomyia, Atrichopogon*) see also Couplet 15b.

Larvae of this family have several forms and they cannot be separated with one set of key characters. The above genera have body forms resembling Figs. 250 and 251, but they are characterized by the presence of false legs on the prothorax *plus* the dorsal side of the body has short but distinct setae. They presumably feed on algae and other plant material.

13b Body without tubercles and setae. **14**

14a Uncommon; larvae with unpaired thoracic and terminal abdominal prolegs which are margined with curved bristles; prothorax with a pair of short respiratory tubes (r) and fan-like setae (s) (Fig. 251) (Solitary Midges). .... **Thaumaleidae**

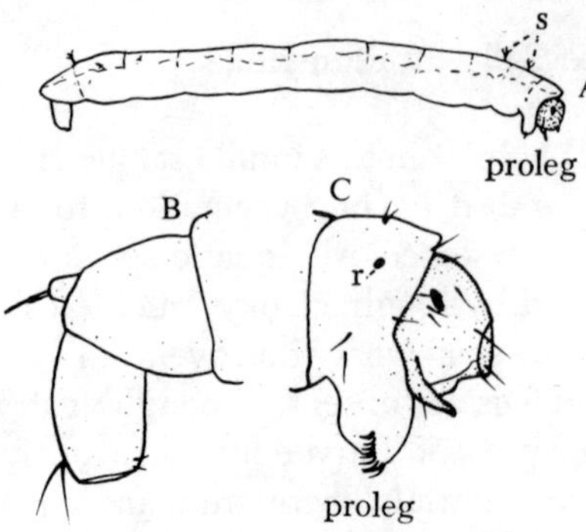

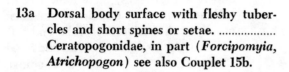

Figure 251

Figure 251 A. Thaumaleidae; B. caudal end; C. Head, S—fan-like setae, r—respiratory openings

These insects have no close relatives but superficially resemble Chironomidae. The thoracic and abdominal segments are sclerotized dorsally. Larvae of these rarely collected flies live in areas where a thin layer of water flows over rocks or through moss. The dorsal side of the larva is often exposed above water, and they are often found on vertical surfaces. They apparently feed on diatoms and other organic material. They are like an odd chironomid in appearance, but differ in having only single anterior and posterior prolegs, as well as in other details mentioned above.

14b Common; larvae usually with paired prolegs; respiratory tubes absent on prothorax (Fig. 252) (Non-Biting Midges). .... **Chironomidae**

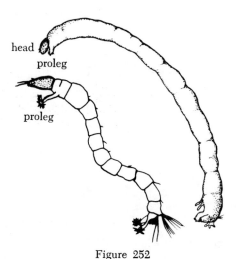

head

proleg

proleg

Figure 252

Figure 252 Chironomidae

Ecologically, this is an extremely important group, and they are often the dominant benthic fauna of lakes. While Tipulidae is the largest family of Diptera, Chironomidae is the largest aquatic family. Unfortunately, due to the large number of species, they are often not identified beyond family or subfamily in aquatic studies. Members of this family are found in almost every conceivable aquatic habitat. Some are phoretic on mayflies and stoneflies. The group includes carnivores, herbivores, case builders, net spinners, and free-living forms. Many have haemoglobin and are bright red in life. Some can live in highly polluted conditions, while others are restricted to cool clear water.

15a End of abdomen with three pairs of long filamentous gills and two false legs (Fig. 253). .... Tanyderidae

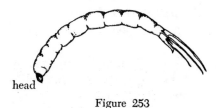

head

Figure 253

Figure 253 Tanyderidae

These are considered to be among the rarest of Diptera. Larvae are about 15-17 mm long when mature. They are readily recognized by the tracheal gills which arise from abdominal segments 8 and 9 and from the pseudopods at the end of the abdomen. Two genera and four species are known from North America. *Protoplasa fitchii* is found along the Atlantic coast while three species of *Protanyderus* occur in the west with records from Alberta to Colorado and west to California. Larvae feed on vegetable matter and burrow in gravel in shallow, fast-flowing water. The long gills often project above the substrate. The rarity of these insects may be largely due to the failure of collectors to search beneath the substrate of gravel floodplains.

15b Tip of abdomen without filaments; with a single retractile false leg or none (Fig. 254) (Punkies, Sandflies, No-see-ums). .. .... Ceratopogonidae, in part (*Dasyhelea, Culicoides, Bezzia, Palpomyia*)

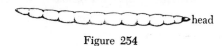

head

Figure 254

Figure 254 Ceratopogonidae

See also Couplet 13a. In these genera dorsal setae are few or absent and anterior false legs are lacking. *Dasyhelia* has a retractile posterior false leg with a row of hooks. *Culicoides, Bezzia* and *Palpomyia* are slender eel-like larvae totally lacking false legs, as in Fig. 254. Larvae are both predatory and herbivorous in habit, depending on the genus.

16a (1b) **Head capsule sclerotized and present but is partly or completely withdrawn into thorax and thus not visible in preserved specimens (Fig. 255); mandibles close toward each other; caudal end of abdomen with fleshly lobes (see also Fig. 249).** ....
.... **(Return to Couplet 10b) Tipulidae**

head

Figure 255

Figure 255 Tipulidae, dissected to show withdrawn head capsule

16b Sclerotized head capsule much reduced; "head" represented by a muscular bundle of sclerotized rods or hooks, all of which are usually not visible in preserved undissected specimens; mandibles parallel (see Figs. 239, 259, 260). ....
.... 17

17a **Larva with a "forked tail" which is fringed with hairs; abdominal segments each with a pair of prolegs which have numerous hooklets; abdominal segments with lateral and dorsal projections (Fig. 256) (Snipe Flies). .... Rhagionidae (= Athericidae)**

head

Figure 256

Figure 256 Rhagionidae

Larvae of the widespread and often common genus *Atherix* inhabit streams and rivers where they are active predators. They are usually easily recognized. Larvae of Ephydridae and some Empididae that may have a forked terminal process lack a fringe of hairs and the body is not as in Fig. 256.

17b Larva without a "forked tail" fringed with hairs; prolegs on abdominal segments not as described. 18

18a **Larva cylindrical and pointed at both ends and with a girdle of false legs completely encircling each segment (arrow, Fig. 257) (Deer and Horse Flies).** ....
.... **Tabanidae**

head

Figure 257

Figure 257 Tabanidae; arrow—false leg

Immature stages of many species are aquatic and they are predatory on other aquatic organisms. The body of the larva is cylindrical and consists of 11 segments and a retractile head. The abdominal segments are ringed with false legs.

18b Larva without a girdle of false legs on each segment. 19

19a Larva with a single long and retractile or short and non-retractile breathing tube at the end of the abdomen (arrow, Fig. 258); ventrally with 7 pairs of prolegs (Flower flies). .... Syrphidae

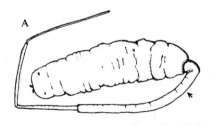

A

B
head

Figure 258

Figure 258 Syrphidae; arrows—breathing tubes. A. *Eristalis* type; B. *Syritta* type

Most larvae of this large family are terrestrial, but many genera and species are truly aquatic. They are characterized by the caudal respiratory organ which may be short and strongly sclerotized, Fig. 258B (near *Syritta*), or very long, as in the common "rat-tailed maggot" (*Eristalis,* Fig. 258A). Many of the long-tailed forms inhabit filthy situations such as outflows from dairy barns and similar naturally occurring habitats, where they are able to breathe by means of the respiratory tube; the *Syritta*-like larvae in Fig. 258B was collected in shallow water in a stony substrate of a rapid and unpolluted river. Larvae feed on organic material as far as is known.

19b Without a breathing tube and 7 pairs of prolegs. 20

20a Mandibles sickle-shaped, often in a re-curved position; "head capsule" usually with definite antennae (Fig. 259). 21

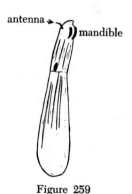

antenna →   mandible

Figure 259

Figure 259 "Head capsule" (Rhagionidae)

20b Mandibles hook-like; "head capsule" reduced to a few sclerotized fragments; antennae absent (Fig. 260). 22

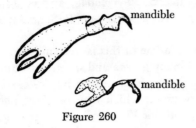

Figure 260

Figure 260   Sclerotized head parts (Ephydridae)

21a   Caudal end terminating in a spiracular pit surrounded by several pointed lobes (Fig. 261) (Long-legged Flies). .... **Dolichopodidae**

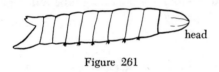

Figure 261

Figure 261   Dolichopodidae

Abdominal segments in these larvae have "creeping ridges" which are armed with chitinous projections. Larvae are usually white and about 10 mm long. This is a large, common, but poorly known family. There are relatively few aquatic members. Aquatic larvae are usually found in mud or decaying vegetation, and they are predatory.

21b   Caudal end rounded with the spiracles borne on the surface or on distinct raised processes (Fig. 262) (Dance Flies). .... **Empididae**

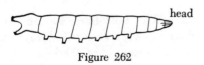

Figure 262

Figure 262   Empididae

There are many terrestrial species in this large family, but about 2 dozen genera in two subfamilies are strictly aquatic. The larvae are predatory and in many ways resemble Dolichopodidae. They differ by lacking lobes at the caudal end but instead have tapered, tail-like processes. Well developed and large ventral pseudopods are present and general appearance is similar to Fig. 262. Most aquatic larvae have been taken from mud or moss in streams.

22a   Larva with a lobed area (Figs. 263B, C) and spiracles at the end of the abdomen (Marsh Flies). .... **Sciomyzidae**

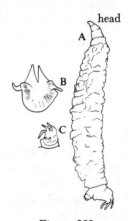

Figure 263

Figure 263   A. Sciomyzidae; B, C. views of caudal end

These have rather tipulid-like lobes surrounding the spiracular disk at the caudal end. They lack the head capsule of Tipulidae and lack the antennae of Dolichopodidae. Larvae often are rather leathery and yellow or brown and are covered with tubercles. They are found in a variety of aquatic habitats where they feed on snails.

22b End of abdomen without such a spiracle-bearing disk. 23

23a Larvae with the end of the abdomen enlarged; living in pitcher plants of the genus *Sarracenia* (Flesh Flies). ....
.... **Sarcophagidae**

While basically terrestrial, a few species in this family inhabit the water in pitcher plants where they feed on the remains of dead insects. They have been reported only in the east. Larvae resemble Fig. 263; in some larvae the end of the abdomen is enlarged to form a cup-like float which bears slit-like spiracular openings.

23b Larva not with end of abdomen enlarged and not living in pitcher plants. 24

24a Tip of abdomen with fleshy bifurcations or a tube which bears the spiracles, often associated with salt lakes or brine pools; integument often with spinules (Fig. 264) (Shore and Brine Flies). ....
.... **Ephydridae**

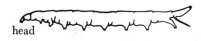

Figure 264

Figure 264   Ephydridae

Larvae are diverse in form. Habitat ranges from fresh water to salt or alkaline water and brine pools. They are common and widespread, with many species. Prolegs may be absent or there may be up to 8 pair, bearing strong hooks. Caudal respiratory organs may be paired tubercles or elongate tubes which can be retracted. Larvae are sometimes so abundant that balls of larvae and pupae can

be seen floating through the water. They feed on algae and similar plant material.

24b Tip of abdomen without fleshy bifurcations or a tube which bears spiracles (Fig. 265) (Housefly group). ....
.... **Muscidae**

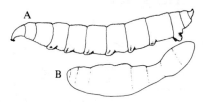

Figure 265

Figure 265   A. Muscidae larva; B. pupa of *Mydaeina*, an arctic form

These larvae are smooth and tapered and are the typical "maggot". Muscids are mainly terrestrial, including such members as the housefly and stable fly. A few are truly aquatic where they are predatory on other organisms. One aquatic form with a well adapted floating pupa is *Mydaeina*, known from Arctic regions (Fig. 265B).

DIPTERA ADULTS

1a Antennae consisting of 6 or more distinct segments; antennae usually (but not always) longer than the thorax and sometimes densely covered with hairs (Figs. 266, 275, 278) (Suborder Nematocera). 2

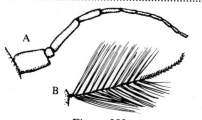

Figure 266

Figure 266   Antennae, A. Tipulidae; B. Culicidae

1b   Antennae shorter than the thorax and appearing to consist of five or fewer segments; the third often bears a bristle-like arista (ar, Figs. 267A and C) or the third may be divided into false segments (Figs. 267B, D; also 289). 14

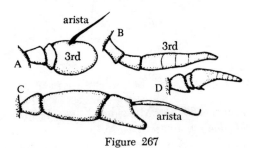

Figure 267

Figure 267   Antennae, A. Syrphidae; B. Stratiomyidae; C. Sciomyzidae; D. Tabanidae

These "higher Diptera" (Suborders Brachycera and Cyclorrhapha), have relatively few aquatic members in total, but a number of families are represented.

2a   Body hairy and moth-like, under 4 mm long; wings hairy, broadly ovate (Fig. 268) (Moth Flies and Sand Flies). .... **Psychodidae**

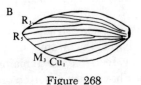

Figure 268

Figure 268   A. Psychodidae; B. wing

These tiny flies resemble their namesake because of the broad hairy wings which are held roof-like over the body when at rest. The wings are bent at the base and fold closely to the body. The antennae are usually about three times the length of the head and made up of bead-like segments. Adults frequent shady wooded areas but are sometimes seen at lights or in house windows. They run and fly with rapid, jerky movements. Some are able to breed in drain pipes of houses, the larvae tolerating soap and hot water. Female *Phlebotomus* (this genus is not truly aquatic) are blood suckers and in tropical areas are sometimes disease vectors. The other 10 genera are harmless, but the food habits of adults are poorly known. There are about 100 North American species. The family is common.

2b   Not hairy and moth-like; size variable; wing not as in Fig. 268B. 3

3a   Small rare flies, body under 3 mm, wings fan-like and under 5 mm; antennae very long in the male and inserted above the eye in both sexes (Fig. 269) (Mountain Midges). .... **Deuterophlebiidae**

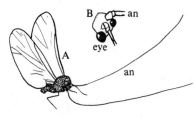

Figure 269

Figure 269 A. Deuterophlebiidae male; B. head; an—antenna

Wings of these flies have net-like fold marks and are broadest in the basal 1/4. The antennae are extremely long in the male but short in the female. Mouthparts are lacking. They are rarely collected and are represented in North America by 4 species, recorded from Alberta to California and east to Wyoming and Colorado. There is but one genus, *Deuterophlebia*, which is also found on the mainland of Asia and Japan. They are found near mountain streams.

3b   Not exactly as figured and described above. 4

4a   Mesonotum with a dorsal "v" or "y" shaped transverse suture, the arms of which originate anterior to the wing bases and with the apex pointed posteriorly (Fig. 270, arrow); at least 9 veins reach the wing margin; legs long and slender, ocelli absent. 5

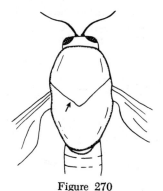

Figure 270

Figure 270   Transverse   suture   of   thorax (Tipulidae)

4b   Either transverse suture of mesonotum incomplete (Fig. 271A) or mesonotum crossed by a straight transverse suture, or suture absent (Fig. 271B); legs variable; usually fewer than 9 veins reach the wing margin; ocelli present or absent. 7

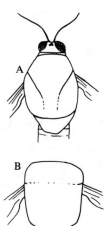

Figure 271

Figure 271   A. Incomplete transverse suture (Blephariceridae); B. straight suture (Muscidae)

5a Wings with two anal veins reaching the margin (veins 2A and 3A, Fig. 272B); common (Craneflies). .... Tipulidae

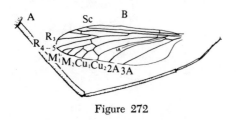

Figure 272

Figure 272   Tipulidae, A. leg; B. wing

This is the largest family in the Order Diptera, but most are not aquatic. There are about 60 North American genera and nearly 1500 species. In the world fauna there are over 13,000 species, and the genera *Tipula* and *Limonia* contain over 1000 species each. Adults are found in shady or humid areas and often are seen at lights. They have very long legs and resemble large mosquitoes. Some species are more than an inch in body length but some are minute. These flies do not bite and they probably take no food other than nectar.

5b Wings with one anal vein reaching the margin (vein 3A, Fig. 273); less common. 6

6a Anal area of wing not expanded (arrow, Fig. 273), hind margin of wing a smooth arc; radius 4 branched (False or Phantom Crane Flies). .... Ptychopteridae

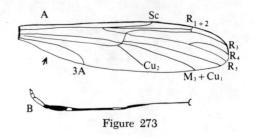

Figure 273

Figure 273   Ptychopteridae, A. wing; B. leg

While resembling craneflies, these are considered to be more closely related to the Psychodidae. The V suture of the dorsal thorax charcteristic of the tipulids is sometimes faint and is often Y-shaped. Ocelli are absent. The family is small and relatively uncommon, but widespread, with three genera and less than 30 species. *Ptychcptera* spp. adults resemble large inconspicuous gnats, but *Bittacomorpha* and *Bittacomorphella* have striking inky black and pure white banded legs (Fig. 273B). Though the body is small, the legs spread more than an inch as these beautiful flies drift in the wind.

6b Anal area of wing expanded, hind margin distinctly angular, similar to Fig. 274 (arrow); radius 5 branched (Primitive Craneflies). .... Tanyderidae

Protoplasa fitchii and three species of *Protanyderus* make up this family in North America. They are widespread in North America but are very rarely seen. While resembling crane flies they are considered to be closer to the Psychodidae. The wings usually have a banded pattern. Ocelli are absent. Body length is from 5 to over 10 mm.

7a Legs long and slender, cranefly-like; transverse suture incomplete as in Fig. 271A; wings sometimes with a spider-web-like network of creases superimposed upon the primary venation; anal area cf wing expanded to a rounded point (arrow, Fig. 274B) (Net-winged Midges). .... Blephariceridae

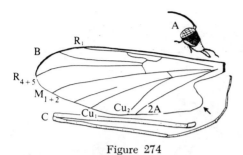

Figure 274

Figure 274 Blephariceridae, A. Side view of head; B. wing; C. leg

These half-inch long flies resemble craneflies (Tipulidae) except that here ocelli are present on a protuberance on the top of the head between the eyes and the transverse suture on the thorax is interrupted rather than a y- or v-shaped suture (Fig. 271A). The eyes are divided into an upper light and lower black portion. The mouthparts are extended and females are predacious. There are 5 North American genera and about 2 dozen species. They are associated with rapid streams in both eastern and western North America.

7b   Wings without a network of creases; legs not extremely long and cranefly-like, *or* anal area of wing not expanded. 8

8a   Wings very broad with anterior veins thick, other veins weak (Fig. 275A); hump-backed flies usually 3mm or less; antenna about as long as head (Blackflies). .... Simuliidae

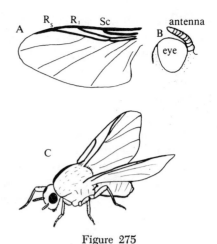

Figure 275

Figure 275 Simuliidae, A. wing; B. head and antenna; C. body shape (diagrammatic)

These are small, robust, dark colored flies with short legs and a hump-backed appearance. Females of many species are blood feeders and attack warm-blooded animals. Many bite man. Dense swarms are known to have caused death in domestic animals. The family has a worldwide distribution, often being very abundant. Tropical species often transmit disease, such as Onchoceriasis (river blindness).

8b   Wing narrower; if anterior veins are thicker than the posterior veins, then not hump-backed nor with short antenna (Midges). 9

9a   Costa ending at or before the wing tip (Figs. 276, 277). 10

The thickened edge ends at or before the wing tip. This couplet may be unclear for some specimens, but no alternative has been discovered. Specimens keyed to this point may require careful comparisons with the drawings and descriptions of families in couplets 10-13.

9b Costa continuing around wing tip (Figs. 278, 279), i.e., thickness of edge of wing is uniform around the wing tip. 11

10a Head spherical and not hidden in dorsal view by the thorax (th; Fig. 276A); dorsal thorax rounded and without a median longitudinal ridge or keel; mouthparts for piercing (Sandflies, punkies, no-see-ums). ....
.... Ceratopogonidae

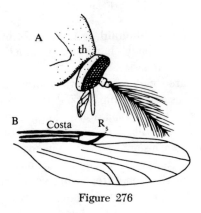

Figure 276

Figure 276 Ceratopogonidae, A. Head region; B. wing

These are small to minute flies, commonly 1-2 mm but up to 5 mm long that are nasty biters. The legs are of moderate length and the hind pair are usually longest. Antennae of males are plumose. They were formerly considered to be part of the family Chironomidae. Some adults feed on blood of vertebrates while other species attack insects, especially midges, mosquitoes, and dragonflies and their blood is sucked out. Some are known to be vectors of diseases such as filariasis. They are widespread and common, with about 400 aquatic species.

10b Head somewhat flattened and partly hidden in dorsal view by the thorax (th, Fig. 277B); metanotum and meso-

notum (dorsal thorax) with a distinct median longitudinal keel or ridge; mouthparts not for piercing (Non-biting midges). .... Chironomidae

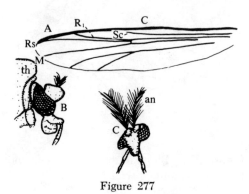

Figure 277

Figure 277 Chironomidae, A. Wing; B. head and thorax (th); C. front view of head with antenna (an)

While many species are small, some are over 10 mm in length. They do not bite but are often an annoyance because of huge emergences from eutrophic lakes or ponds, the insects covering the screens of lighted windows. The inexperienced often mistake them for mosquitoes. In life, midges characteristically rest with the front legs raised in the air, but mosquitoes raise the hind legs. These are the most common aquatic Diptera (about 1000 species). Antennae of males are plumose (Fig. 277C).

11a Seven longitudinal veins reaching wing margin; rare (Solitary Midges). ....
.... Thaumaleidae....

These are small, stocky, shining yellow to brown flies, 3 to 4 mm in length, which have a small head and a short antenna consisting of two globose basal segments and a rather bristle-like flagellum consisting of 10 obscure segments. The palpi are longer than the antennae, the eyes are quite large and cover most of the head in both sexes, and ocelli are lacking.

Wings often bend sharply at the base in pinned specimens, folding roof-like as in Psychodidae. Little is known of the biology of adults except that they are found in vegetation not far from the larval habitat. Two genera and 4 species are known for N.A., where they occur in both the east and west in mountainous areas.

11b   At least 9 veins reach wing margin (Fig. 278); common. 12

12a   In outer 1/5 of wing, veins R_1 - R_4 distinctly arched and curved to the rear (Fig. 278) (Dixid Midges). ...... Dixidae

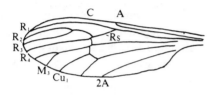

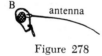

Figure 278

Figure 278   Dixidae, A. Wing; B. head and non-plumose antenna

Dixidae are small, nearly bare midges with slender legs. They are probably most closely related to the mosquitoes but they do not feed on blood. Adults lack ocelli, have non-plumose 16-segmented antennae, and eyes spaced widely apart in both sexes. The curved R veins and the lack of body scales usually will identify this family. Adults are frequently found in small swarms near damp areas or resting in the shade near water. They are common and widespread.

12b   In outer 1/5 of wing, R veins straight or nearly so, not distinctly curving to the rear (Fig. 279). 13

13a   Proboscis (p) elongate; scales present on wing veins and on the body (Fig. 279) (Mosquitoes). .... Culicidae

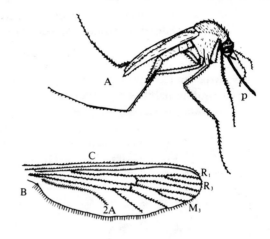

Figure 279

Figure 279   A. Culicidae; B. wing; p-proboscis

Mosquitoes are one of the unfortunate productions of nature which make life miserable in northern and arctic regions as well as many other areas, especially the tropics where disease transmission is a problem. While Chaoboridae (and a few Culicidae) are predatory in the larval stage and have harmless adults, most mosquito larvae feed on a poor diet of organic debris and bacteria and female adults make up for it by requiring a blood meal from a vertebrate prior to egg production. Man is often the vertebrate host.

Mosquitoes are among the most thoroughly studied insects because of the "pest" status. There are about 11 genera and 150 species in N.A. The scales on the body are the most characteristic feature.

13b   Proboscis not elongate; veins and wing margins hairy; scales, if present, restricted to wing margins (Phantom Midges). .... Chaoboridae

This group was formerly considered to belong in the mosquito family but now has separate family status. Chaoboridae lack piercing mouthparts of mosquitoes and have hairy rather than scaled wing veins. Wing venation is similar to Culicidae. Adults are 4-10 mm in length and sometimes have mottled wings; they are pale yellow to brown in color. There are fewer than 20 species in N.A. in 4 genera. Adult feeding habits are unknown.

14a (1b) **Three pads of nearly equal size under the tarsal claws (Fig. 280). 15**

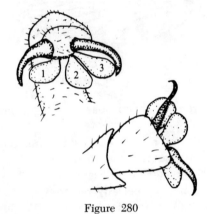

Figure 280

Figure 280 Tarsal claws and pads (numbered)

14b **With two pads separated by a bristle (arrow) under the tarsal claws (Fig. 281). 17**

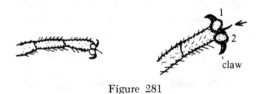

Figure 281

Figure 281 Tarsal claws and pads (numbered)

15a **Third antennal segment subdivided into 3-8 false segments (Fig. 282). 16**

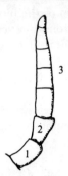

Figure 282

Figure 282 Antenna, Stratiomyidae

15b **Third antennal segment not divided into false segments, third segment with an elongate bristle-like arista (ar, Fig. 283) (Snipe Flies). .... Rhagionidae**

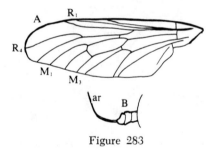

Figure 283

Figure 283 Rhagionidae (not *Atherix*), A. Wing; B. antenna; ar—arista

Only the widespread genus *Atherix* (sometimes in a separate family, Athericidae) is aquatic in this family. *A. pachypus* is known from Washington state and *A. variegata* ranges from the Northwest Territories to Quebec and south to Ga. and California. The often repeated but scarcely believable accounts of egg laying in *Atherix* (masses of dead adult bodies and eggs on objects overhanging water) has been verified by Peter Mason (pers. comm.). Both adults and larvae are predatory. Adults are usually found in moist shady areas. Rhagionids have 3 tarsal pads and the antennae are 3 segmented with the 3rd bearing an arista as long as the 3 segments combined. The arista is short in *Atherix*.

16a Spurs present on tibia of middle leg; costa continuing around tip of wing (arrow, Fig. 284) (Deer and Horseflies). .... **Tabanidae**

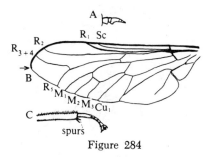

Figure 284   Tabanidae, A. Antenna; B. wing; C. tibia and tarsus

These are biting flies. Adults are stout and without bristles, the third antennal segment is divided into subsegments by external annulations, and the eyes are large. The mouthparts are developed into a proboscis for piercing the skin of animals. In life the eyes are often irridescent and strikingly marked. Empodium and pulvilli are padlike, i.e., there are 3 pads under the claws.

16b Tibial spurs absent; costa ending before wing tip (Fig. 285) (Soldier Flies). .... **Stratiomyidae**

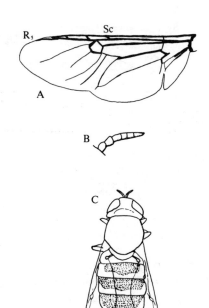

Figure 285

Figure 285   Stratiomyidae, A. Wing; B. antenna; C. body form

These are called soldier flies because many species have striking black and yellow or green color patterns. They lack bristles but they may be covered with a soft pubescence. Species vary considerably in appearance, but the wing venation is fairly constant—the anterior three or more veins are heavy and the posterior veins weak. Adults are often found near flowers. They are common and widespread.

17a Crossvein r-m in midwing crossed by a longitudinal spurious vein which courses between R_{4+5} and M_{1+2} (Fig. 286) (Flower Flies). .... **Syrphidae**

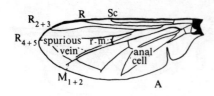

Figure 286

Figure 286 Syrphidae, A. Wing; B. antenna

This group contains many terrestrial species and is widespread and common. While general appearance varies, they are often brightly marked with yellow and black and are immediately recognized by the spurious vein (a longitudinal fold in the wing membrane, not a true vein). They are important as pollinators of plants, and many mimic Hymenoptera. There are numerous species and about 6 genera of aquatic Syrphidae.

17b Without a spurious vein between R_{4+5} and M_{1+2}. 18

18a Frontal suture entirely absent, front of head uniformly sclerotized (Fig. 287); no alula (white or clear flap at base of wing). 19

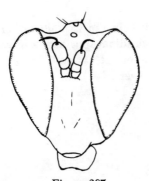

Figure 287

Figure 287 Dolichopodidae, front view of head

18b Frontal suture well developed as a horseshoe-shaped groove over the antennae, continuing down to separate the center of the face from the sides (arrows, Fig. 288). 20

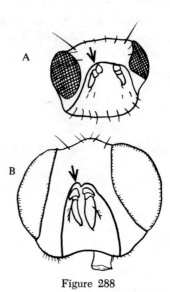

Figure 288

Figure 288 A. Ephydridae and B. Muscidae, with frontal suture (arrows)

19a Crossvein r-m located in basal 1/4 of wing (Fig. 289B) (Long-legged Flies). .... Dolichopodidae

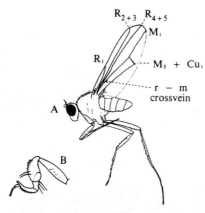

Figure 289

Figure 289 Dolichopodidae, A. Entire insect; B. leg of male

This is a large family of medium to small, metallic green, long-legged flies. Legs of males are often modified. Adults engage in mating dances and males may have modified wings, legs, or antennae. Both adults and larvae are predatory, adults feeding on midges or other small insects. Adults are abundant near water on mud flats or over sand; many skate on the water surface and are difficult to catch. They are common and widespread.

19b   Crossvein r-m located beyond the basal 1/4 of the wing (Fig. 290) (Dance Flies). .... **Empididae**

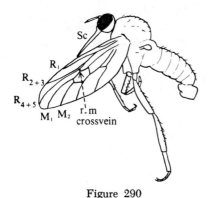

Figure 290

Figure 290   Empididae

This is a large family of flies which are mostly under 10 mm in length. The head appears spherical and loosely attached to the body, giving the appearance of a neck. The proboscis is often well developed and rigid. The body is usually somewhat hairy and dull grey or brown, but never metallic. Male genitalia are often large. They engage in elaborate mating swams and dances, and males offer captured prey or a silvery "balloon" to females.

20a   **A large white or clear flap (alula) at base of wing (arrow, Fig. 291); transverse suture (Fig. 271B) extends across thorax.** **21**

These are the calypterate muscoid flies. Sometimes the two families that follow are combined in th family Anthomyiidae.

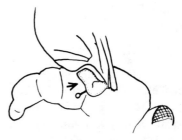

Figure 291

Figure 291   Alula (arrow)

20b   **Flap at the base of wing small or absent; transverse suture incomplete.** **22**

21a   **Vein M_{1+2} bending forward, narrowing the R_5 cell at the wing margin (Fig. 292) (Flesh Flies).** .... **Sarcophagidae**

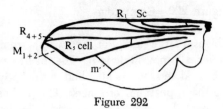

Figure 292

Figure 292 Sarcophagidae

This group includes parasites of insects and scavengers as well as flesh flies. They are basically terrestrial, but several species of *Sarcophaga* have larvae that live in water in pitcher plants. Adults feed on nectar and plant juices. Members of the family are generally greyish with longitudinal thoracic stripes and a nonmetallic, bristly body.

21b   Vein M_{1+2} nearly straight, R_5 cell not narrowed near wing margin. ....
.... **Anthomyiidae**

This group is often included in the Muscidae or housefly family (see larval key). They are primarily terrestrial but a few are true aquatics.

22a   Costa (C) of the wing broken in two places, at the humeral crossvein (H) and near the end of the subcosta (Sc); sub-costa incomplete and not continuing to the costa (Fig. 293) (Shore and Brine Flies). .... **Ephydridae**

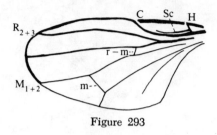

Figure 293

Figure 293 Ephydridae

These are very common, small to minute, dull colored flies which have the lower part of the head cut away giving the appearance of a gaping mouth. They often occur on shores walking over mud and water feeding on diatoms and algae.

22b   Costa not broken; subcosta complete and ending at the costa (Fig. 294) (Marsh Flies). .... Sciomyzidae

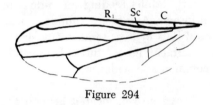

Figure 294

Figure 294 Sciomyzidae

In addition to the key characters, living adults are recognized by the antennae which are held forward horizontally. Many are brown or grey but a few are reddish. Many have spots on the wings. They are common and often seen sitting on plant stems or other upright objects near the larval habitat.

Appendix I

Orders and Families with Common Names and Synonyms

I. **Collembola**–Springtails
Entomobryidae
Hypogastruridae
Isotomidae
Onychiuridae
Poduridae
Sminthuridae

II. **Ephemeroptera**–Mayflies
Ametropodidae
Baetidae–Small Mayflies
Baetiscidae
Behningiidae
Caenidae
Ephemerellidae
Ephemeridae (includes Palingeniidae)–
Burrowing Mayflies
Heptageniidae–Stream Mayflies
Leptophlebiidae
Metretopodidae
Neoephemeridae
Oligoneuriidae
Polymitarcyidae
Potamanthidae
Siphlonuridae
Tricorythidae

III. **Odonata**–Dragonflies and Damselflies
Aeshnidae=Aeschnidae–Darners
Agrionidae=Agriidae=Calopterygidae–
Broad-winged Damselflies
Coenagrionidae=Coenagriidae–Narrow-
winged Damselflies
Cordulegastridae–Biddies
Gomphidae–Clubtails
Lestidae–Spread-winged Damselflies
Libellulidae–Common Skimmers

Petaluridae–Graybacks
Protoneuridae

IV. **Orthoptera**–Grasshoppers and Crickets

V. **Plecoptera**–Stoneflies
Capniidae
Chloroperlidae–Green Stoneflies
Leuctridae
Nemouridae–Spring Stoneflies
Peltoperlidae–Roachlike Stoneflies
Perlidae–Common Stoneflies
Perlodidae–Perlodid Stoneflies
Pteronarcidae–Giant Stoneflies
Taeniopterygidae

VI. **Hemiptera**–True Bugs
Belostomatidae–Giant Water Bugs
Corixidae–Water Boatmen
Dipsocoridae–Jumping Ground Bugs
Gelastocoridae–Toad Bugs
Gerridae–Water Striders
Hebridae–Velvet Water Bugs
Hydrometridae–Water Measurers
Macroveliidae
Mesoveliidae–Water Treaders
Naucoridae–Creeping Water Bugs
Nepidae–Water Scorpions
Notonectidae–Back Swimmers
Ochteridae
Pleidae–Small Backswimmers
Saldidae–Shore Bugs
Veliidae–Ripple Bugs

VII. **Megaloptera**–Dobsonflies, Alderflies
Corydalidae–Hellgrammites, Dobsonflies
Sialidae–Alderflies

VIII. **Neuroptera**–Lacewings
Sisyridae–Spongillaflies

IX. **Trichoptera**—Caddisflies
    Beraeidae
    Brachycentridae
    Calamoceratidae
    Glossosomatidae
    Helicopsychidae—Snail-case Caddisflies
    Hydropsychidae—Net-spinning Caddisflies
    Hydroptilidae—Micro-caddisflies
    Lepidostomatidae
    Leptoceridae—Long-horned Caddisflies
    Limnephilidae—Northern Caddisflies
    Molannidae
    Odontoceridae
    Philopotamidae—Finger-net Caddisflies
    Phryganeidae—Large Caddisflies
    Polycentropodidae—Tube-making Caddisflies
    Psychomyiidae
    Rhyacophilidae—Primitive Caddisflies
    Sericostomatidae

X. **Lepiodoptera**—Moths
    Pyralidae—Pyralid or Snout Moths

XI. **Coleoptera**—Beetles
    Amphizoidae—Trout-stream Beetles
    Carabidae—Ground Beetles
    Chrysomelidae—Leaf Beetles
    Curculionidae—Snout Beetles
    Dryopidae—Long-toed Water Beetles
    Dytiscidae—Predaceous Diving Beetles
    Elmidae—Riffle Beetles
    Eurystethidae=Salpingidae
    Gyrinidae—Whirlygig Beetles
    Haliplidae—Crawling Water Beetles
    Helodidae—Marsh Beetles
    Heteroceridae
    Hydraenidae
    Hydrophilidae—Water Scavenger Beetles
    Hydroscaphidae

    Limnichidae
    Melyridae
    Psephenidae—Water-penny Beetles
    Ptilodactylidae
    Sphaeriidae
    Staphylinidae—Rove Beetles

XII. **Diptera**—True Flies
    Blephariceridae—Net-winged Midges
    Ceratopogonidae=Heleidae—Sand Flies, Punkies, No-see-ums, Biting Midges
    Chaoboridae (formerly in Culicidae)—Phantom Midges
    Chironomidae=(Tendipedidae)—Non-biting Midges
    Culicidae (restricted sense)—Mosquitoes
    Deuterophlebiidae—Mountain Midges
    Dixidae—Dixid Midges
    Dolichopodidae—Long-legged Flies
    Empididae=Empidae—Dance Flies
    Ephydridae—Shore and Brine Flies
    Muscidae (includes Anthomyiidae)—Housefly group
    Psychodidae—Moth Flies and Sand Flies
    Ptychopteridae=Liriopeidae—False or Phantom Craneflies
    Rhagionidae=Athericidae=Leptidae—Snipe Flies
    Sarcophagidae—Flesh Flies
    Sciomyzidae=Tetanoceridae—Marsh Flies
    Simuliidae—Black Flies or Buffalo Gnats
    Stratiomyidae—Soldier Flies
    Syrphidae—Flower Flies
    Tabanidae—Horse and Deer Flies
    Tanyderidae—Primitive Craneflies
    Thaumaleidae=Orphnephilidae—Solitary Midges
    Tipulidae—Common Craneflies

XIII. **Hymenoptera**—Wasps

Appendix II

Selected Journals and Societies

Eatonia, A Newsletter for Ephemeropterists
Janice G. Peters, Editor
School of Science and Technology
Florida A and M University
Tallahassee, Florida

Entomological Society of America
(Annals of)
4603 Calvert Road
College Park, Maryland 20740

Entomological Society of Canada
(The Canadian Entomologist)
Treasurer
1320 Carling Ave.
Ottawa, Ontario
K1Z 7K9

The Entomological Society of Washington
(Proceedings of)
Dept. of Entomology
Smithsonian Institution
Washington, D.C. 20560

International Odonatological Society
(Odonatologica)
Dr. J. M. van Brink, Treasurer
Institute of Genetics
University of Utrecht
Transitorium III, Padualaan 8
Utrecht, the Netherlands

Kansas Entomological Society
(Journal of)
Secretary-Treasurer
Kansas State University
Manhattan, Kansas

North American Benthological Society
Secretary
Coop Fishery Research Unit
Stephens Hall, University of Missouri
Columbia, Missouri 65201

The Pacific Coast Entomological Society
(The Pan-Pacific Entomologist)
The Secretary, Vector Control Section
California Dept. of Health
2151 Berkeley Way
Berkeley, California 94704

Perla, A Newsletter for Plecopterologists
R. W. Bauman
Dept. of Zoology
Brigham Young University
Provo, Utah 84602

TIEG (Teen International Entomology Group)
TIEG Magazine
Dept. of Entomology
Michigan State University
East Lansing, Michigan 48824

BIOLOGICAL SUPPLY HOUSES

AMBI
1330 Dillon Heights Avenue
Baltimore, Maryland 21228
U.S.A.

BioQuip Products
P.O. Box 61
Santa Monica, California 90406
U.S.A.

Entomological Supplies, Inc.
5655 Oregon Avenue
Baltimore, Maryland 21227
U.S.A.

Ward's Natural Science Establishment, Inc.
P.O. Box 1712
Rochester, New York 14603
U.S.A.

Wildco Instruments
301 Coss Street
Saginaw, Michigan 48602
U.S.A.

Index and Glossary